JOHN WORMALD

D1784208

Diffraction methods

Clarendon Press · Oxford · 1973

Oxford University Press, Ely House, London W.1

GLASGOW NEW YORK TORONTO MELBOURNE WELLINGTON
CAPE TOWN IBADAN NAIROBI DAR ES SALAAM LUSAKA ADDIS ABABA
DELHI BOMBAY CALCUTTA MADRAS KARACHI LAHORE DACCA
KUALA LUMPUR SINGAPORE HONG KONG TOKYO

PRINTED IN GREAT BRITAIN BY
J. W. ARROWSMITH LTD., BRISTOL, ENGLAND

Editor's foreword

AMONG the many methods available to determine the structures of chemical compounds diffraction methods occupy a special position for several reasons. They have a long history, yet are in a very active state of development; their range is very wide; the general principles underlying them are simple but the actual interpretation of diffraction patterns is often believed to be beyond the comprehension of ordinary chemists. The most important purpose of this book is therefore to show (with a minimum of mathematical detail) the essential steps whereby a diffraction pattern is translated into a detailed three-dimensional structure. The second purpose is to survey all the major diffraction methods from their early stages to their present sophisticated level and to show by numerous examples the very wide scope of these methods. The third purpose of the book is to indicate how often the rigorous analysis of a diffraction pattern must yield the ultimate answer to a structural problem for which other methods have given solutions which are either uncertain or ambiguous.

There are close connections between this book and others in the series on physical methods, particularly *Magnetic resonance* and the forthcoming volumes on spectroscopic methods. The structures of small molecules are discussed in *The shape and structure of molecules* and *Stereochemistry and mechanism*, while inorganic lattices will be covered in books about the chemistry of solids.

A.K.H.

Preface

CHEMISTRY seeks answers to two general questions: how the elements combine to form the myriad species found in nature or artificially synthesized, and which rules govern these combinations. The species so formed exhibit a wide range of behaviour, from great stability to violent reactivity, and chemistry itself divides roughly into the two areas of structures and reactions. To ascertain the *why* in either domain, we must first know the *how*. Our factual knowledge is gained through experiments, often of an increasingly 'physical' nature. Chemistry is a living and growing subject, and the student needs to balance his grasp of received knowledge with an understanding of how it is acquired. Both kinds of information ultimately come from reading the periodical literature, past and present; as it is very voluminous and often highly specialized, textbooks and review articles play a role of selection and predigestion for less developed tastes. Facts are well covered in a multitude of textbooks for students but methods are generally described in specialized monographs that often favour internal details over a balanced view of applications and scope.

I have sought here to describe a range of physical methods widely used in chemistry to examine chemical and physical structures, based on the diffraction of waves. Although far from being our only source of information, diffraction methods have proved unusually valuable. All the physical sciences perforce use models in discussing the behaviour of nature; those presented here are especially vivid and direct. In fact, they constitute our closest approach to actually 'seeing' chemical structures; much of the factual material of undergraduate courses in inorganic chemistry, for example, rests on them. I have tried to keep a balance between descriptions of the techniques and an assessment of their practical value. So short a book must necessarily suffer from compression, facile descriptions, and errors. I hope, however, that it will provide a first introduction to the subject for first- and second-year undergraduates. It is meant to be quite easy reading and to form part of a comprehensive series of short texts on chemistry, which will help to put diffraction methods into their wider context.

The treatment is therefore deliberately non-mathematical, although some relevant derivations are presented in Appendix II. The historical development of the subject is briefly indicated and is complemented by a few references to current applications. These are not essential reading to understand the book— rather, they are for the interested reader who may wish to sample first-hand accounts of experiments or find more rigorous definitions and derivations. The original literature can be confusing but is also frequently more direct and refreshing than textbook accounts. I have suggested a few textbooks for further reading, as introductions to subjects of general interest, peripheral to diffraction methods. SI units have been used throughout and their equivalents in the older units still current in the literature are given in Appendix I.

I have inevitably drawn on various sources for illustrations and data, and I wish to express my thanks to the authors and publishers listed in the Acknowledgement for their kind permission to reproduce or adapt their material. I am grateful to Professor Melvyn R. Churchill, who originally introduced me to diffraction methods. The completion of the book owes much to the patient support of my wife, who had to live with my involvement in writing during an already crowded year. My thanks also to Mrs. Pat Pearce, who typed the manuscript admirably at short notice.

J. WORMALD

Acknowledgments

Diagrams and tables have been reproduced or adapted with the kind permission of the following authors and publishers.

Figure 2.1, W. J. Moore (1962), *Physical chemistry*, 3rd Edition, Prentice-Hall, Inc., Englewood Cliffs, N.J.; Fig. 3.5(b), U. W. Arndt and B. T. M. Willis, *Single crystal diffractometry*, © 1966 Cambridge University Press; Fig. 3.6, S. A. Bezman, M. R. Churchill, J. A. Osborn, and J. Wormald, *J. Am. chem. Soc.*, **93**, 2063, © 1971, The American Chemical Society; Fig. 4.6, W. E. Addison (1961), *Structural principles in inorganic compounds*, Longmans; Fig. 4.7, A. F. Wells (1962), *Structural inorganic chemistry*, 3rd ed, Clarendon Press, Oxford; Fig. 4.8, E. B. Fleischer, *J. Am. chem. Soc.*, **94**, 1382, © 1972, The American Chemical Society; Fig. 5.7, L. E. Alexander (1969), *X-ray diffraction methods in polymer science*, Wiley-Interscience; Fig. 5.9, R. A. Fava (1971), *J. Polym. Sci. Part D*, **5**, 3, Wiley-Interscience; Fig. 6.1, G. E. Bacon (1966), *X-ray and neutron diffraction*, Pergamon Press Ltd.; Fig. 6.6, M. R. Churchill and J. Wormald, *J. Am. chem. Soc.*, **93**, 5670, © 1971, The American Chemical Society; Fig. 6.11, C. S. Barrett and T. B. Massalski, *Structure of metals*, © 1966, McGraw Hill, Inc.; Fig. 7.3, R. K. Bohn and A. Haaland (1966), *J. organomet. Chem. (Amsterdam)*, **5**, 470; Fig. 8.1, J. M. Corless (1972), *Nature*, **237**, 229.

Contents

1. THE DIFFRACTION PHENOMENON   1

Introduction. The principles of diffraction. X-rays and crystals. The diffraction of particle beams.

2. SINGLE CRYSTALS AND X-RAYS   7

X-ray crystallography. Single crystals. Diffraction from lattices. Atoms in lattices. The phase problem. Extension of phasing. Refinement. Conclusion.

3. PRACTICAL STRUCTURE DETERMINATION   20

The production of X-rays. Determining the unit cell and space group. Intensity data. Data reduction. Elucidating the structure. Crystallographic results.

4. THE IMPORTANCE OF STRUCTURAL ANALYSES   35

The growth of X-ray crystallography. Inorganic chemistry and mineralogy. Metal complexes and covalent molecules. Organometallic compounds. Metals and semiconductors. Organic compounds. Biological structures. Single-crystal work in perpective.

5. SPECIALIZED APPLICATIONS OF X-RAY DIFFRACTION   55

The powder method. X-ray diffraction and high polymers. Degree of crystallinity. Orientation. Micro- and macro-structure in polymers. Other applications of X-ray diffraction.

6. NEUTRON DIFFRACTION   68

The diffraction of thermal neutrons. Elastically scattered neutrons. Magnetic scattering of neutrons. Inelastic neutron scattering.

7. ELECTRON DIFFRACTION   80

Introduction. Electron diffraction from gases and vapours. High-energy electron diffraction from solids. Low-energy electron diffraction.

8. CONCLUSION   89

APPENDIX I   92
Crystallographic units of length.

APPENDIX II   93

The structure-factor equation. Fourier series and the Fourier transform.

BIBLIOGRAPHY   95

ANSWERS TO PROBLEMS   98

INDEX   99

1. The diffraction phenomenon

Introduction

DIFFRACTION is no new phenomenon: it was mentioned by Leonardo da Vinci, although its association with the wave nature of light was first explained by Fresnel in the nineteenth century. Applications to chemical structures had to wait for the discovery of radiation of suitable wavelengths, in the form of X-rays, by Röntgen in 1895. Friedrich, Knipping, and Laue demonstrated X-ray diffraction from single crystals in 1912; their paper reads as a synthesis of the study of electromagnetic radiation and of classical crystallography and lies at the source of the new science of X-ray crystallography. The Braggs, father and son, took it up immediately and clearly showed the relationship between the diffraction pattern and structure of the crystal; in effect, they performed the first crystal-structure determinations.

From these flowed successive applications to simple ionic compounds, confirming the existing notions about their structures and absolutely validating the laws of Dalton and Avogadro. The theories of classical crystallography and the speculations of Haüy on the internal nature of crystals were confirmed. As the techniques were further refined, they were applied to an increasing number of salts and minerals. Their ionic lattices exhibit a high degree of symmetry, consequent upon the isotropic nature of the electrostatic forces that bind the ions into crystals. The chemical and crystal orders are thus intimately associated.

The limited valence properties of carbon, as observed in organic chemistry, enabled Le Bel and van't Hoff to explain optical activity in terms of tetrahedral stereochemistry at carbon atoms. Great strides were made through degradative and synthetic studies, later bolstered by spectroscopic techniques, and organic chemists quickly gained a clear understanding of structural principles, unaided by direct observation of structures. This direct view has, however, become increasingly desirable as the complexity of newly discovered and synthesized organic molecules has increased. Although W. H. Bragg tackled an organic structure as early as 1921, X-ray diffraction is only now becoming competitive for the examination of new compounds. It has, however, confirmed the structural hypotheses of organic chemistry and is more especially valuable when exact details of molecular stereochemistry must be ascertained. This is notably so in the study of molecular biology and of the mechanisms of drug action. X-ray diffraction has extended its reach from the simple salts and metals to such complex structures as enzymes; it has become a scientific 'big business', involving thousands of participants throughout the world, who contribute to a range of applications.

While this has constituted the mainstream of development, there have been parallel advances in diffraction studies on liquids, powders, and gases and in the use of electron and neutron beams to replace X-rays. These techniques have made more specialized but very valuable contributions, ranging from work on heterogeneous catalysts to investigations of the nature of magnetic solids, from chemical fingerprinting to polymer science. In fact, hardly an issue of a major chemical journal appears without containing some publication based on the application of diffraction methods to chemical problems. We shall attempt a brief review of this phenomenon and an assessment of its past, present and future importance. The overriding aim will be to examine diffraction methods as tools for the practising chemist, taking into consideration the overlaps with adjoining disciplines.

The principles of diffraction

A brief review of the physical principles of diffraction is in order, since it will help the reader to appreciate the potential and limitations of the methods that form the subject matter of this book. Textbooks on optics provide a full account of the phenomenon, as observed with visible light; although only a general discussion can be given here, it makes a useful point of departure and two simple experiments will be recalled.

Fresnel predicted that diffraction effects should arise from the edges of a disc; practical confirmation was provided by Arago and the principle involved is shown in Fig. 1.1. Light originating from a point source is propagated through space as a series of electromagnetic waves, radiating from the point as a concentric series of sinusoidal crests and troughs. For simplicity, the crests alone are represented as rings. The edges of the disc reradiate the incident energy, acting as scattering centres; each element of the circumference is in effect a new point source and new series of concentric waves radiate from it. (The effect can be simulated by floating a stick in a calm pond and dropping in a stone off its beam). On the right of the disc in Fig. 1.1, the wavelets recombine, with coincident crests from different sources reinforcing each other, whereas a combination of crest and trough produces zero amplitude. Reinforcement occurs in certain set directions, indicated by *wavefronts*, shown here as tangents to the participating wavelets; the directions, which are normal to these tangents, are controlled by the distance between successive wavelets—i.e. the wavelength of the light—and the diameter of the disc. Fig. 1.1 is, of course, a simplified two-dimensional view; the actual effect seen by the observer on the right is as of an area of light seen at the centre of the shadow of the disc.

This experiment is readily translated into the more general case of Fig. 1.2. Here, a wavefront, represented by AB', is just approaching the object AB, which contains point scatterers A and B separated by a distance d. The

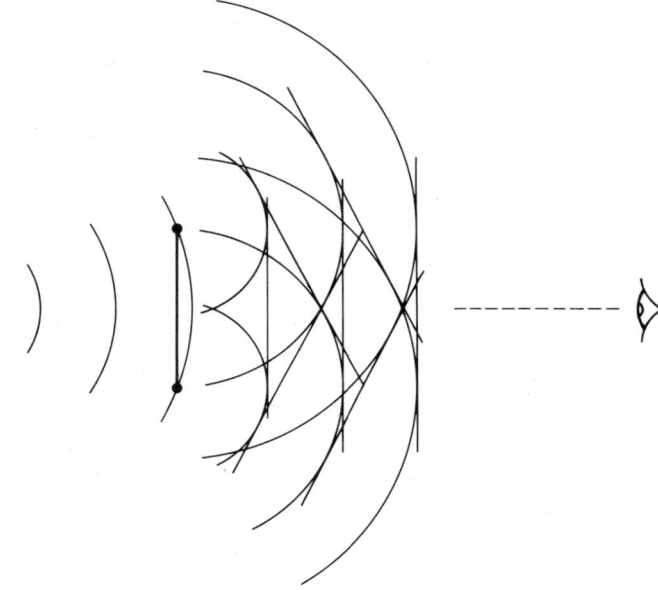

FIG. 1.1. Diffraction of monochromatic light by a plane disc.

scattering of the light of wavelength λ by A and B cooperatively interacts to produce a new departing wavefront, AA'. The angles of incidence and of departure are ϕ and θ, respectively. The path lengths for rays travelling via A and B differ by a distance B'B+BA' whose value is $d(\sin \phi + \sin \theta)$. For cooperative reinforcement to occur the path length difference must equal an integral number of wavelengths. The *diffraction condition* is thus governed by eqn (1.1), in which n is an integer. The wavelength of visible light ranges from

$$n\lambda = d(\sin \phi + \sin \theta) \tag{1.1}$$

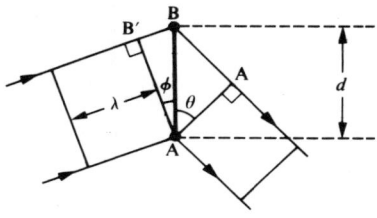

FIG. 1.2. Diffraction from a dipolar point scatterer.

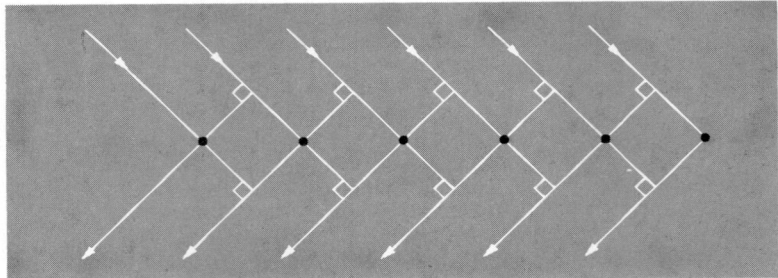

Fɪɢ. 1.3. Cooperative diffraction from an array of points.

about 400 to 800 nm, so that such an experiment with objects of practical size must lead to very small diffraction angles. A diatomic gas molecule is such an object on the molecular scale of dimensions and can be examined with radiation of appropriate wavelength; unfortunately, a gas contains a multitude of such molecules in quite random orientations. This complicates the problem, and diffraction from gases is therefore reserved for Chapter 7.

Our second optical model is the diffraction grating with equally spaced, narrow lines. A transmission grating of this kind is seen in Fig. 1.3, for which it should be clear that the analysis expressed in eqn (1.1) still holds—the several lines merely act cumulatively. We have thus introduced the important concept of cooperative scattering and diffraction by an ordered array of scattering centres. Now imagine a two-dimensional array created by two such gratings superimposed, as in Fig. 1.4. If scattering occurs only at the points lying at the intersections of lines then the diffraction conditions for both gratings must be obeyed simultaneously under the control of eqns (1.2) and

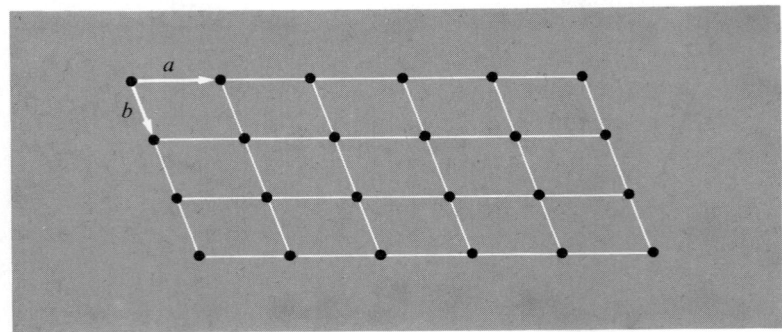

Fɪɢ. 1.4. A two-dimensional array of points.

(1.3). The suffixes a and b refer to the corresponding directions in the array. By a fairly obvious piece of mathematical induction the

$$n_a\lambda = a(\sin \phi_a + \sin \theta_a) \tag{1.2}$$

$$n_b\lambda = b(\sin \phi_b + \sin \theta_b) \tag{1.3}$$

$$n_c\lambda = c(\sin \phi_c + \sin \theta_c) \tag{1.4}$$

extension of the array into a third dimension c will require the addition of eqn (1.4). Without prejudging the nature of crystals, we may point out that this is the form of the triplet of equations derived by Friedrich, Knipping, and Laue (1912). The important principle involved is that of reinforcement from many scattering elements acting in concert. A practical optical demonstration for two dimensions can be had by piercing an aluminium foil with an array of pinholes and illuminating it from a point source of monochromatic light. The diffraction pattern received on a photographic plate is well illustrated by Lipson and Cochran (1966).

X-rays and crystals

In order to produce detectable diffraction patterns from ordered systems of molecular dimensions, the early experimenters required radiation of a wavelength comparable to those dimensions. Laue and his students first realized that X-rays and single crystals of salts might offer a suitable combination and proceeded to test their hypothesis. The known masses of the elements, the densities of copper(II) sulphate pentahydrate and of the zinc blende form of zinc sulphide, and the then-existing notions of crystal structure indicated interatomic spacings of some 100 pm, whereas X-ray wavelengths were known to be of the order of 10 pm. The experimental test was dramatically successful —their paper reads as an almost classic description of the scientific method of hypothesis, test, and deduction—and diffracted beams controlled by the *Laue equations* were received as spots on a photographic plate. This initial success profoundly affected the subsequent development and exploitation of diffraction methods and many of the results of greatest value to chemists have come from work with single crystals. We shall use single-crystal X-ray diffraction in expounding the basic theory and practice of diffraction methods and proceed to other techniques from that base.

The diffraction of particle beams

The wave nature of light has long been accepted and is implicit in the arguments on p. 2. X-rays form a different region of the same spectrum of electromagnetic radiation and are taken to have a similar nature. Since the work of Einstein on photoemission, however, the wave–particle duality in the nature of light has been recognized and quantum mechanics treats this duality as universal. Only the nature of the experiments we perform brings out one or other aspect more strongly. In fact each particle possesses an associated wavelength governed by the de Broglie relationship of eqn (1.5). The

$$\lambda = h/mv \tag{1.5}$$

wavelength λ is related to Planck's constant h, the mass m and the velocity v. An electron with an energy of 100 eV has an associated wavelength of some 120 pm—comparable to molecular dimensions. Whereas the famous experiment of J. J. Thomson with the paddle wheel demonstrated the corpuscular nature of cathode rays, an equally important proof of their wave nature was given by his son who produced diffraction of electrons transmitted through a thin gold foil. A similar experiment with reflected electrons of lower energy was carried out by Davisson and Germer. Electron diffraction is somewhat limited by the relatively low penetrating power of the beams, arising from the charge and low mass of the electron, but has found considerable use in the study of thin specimens, adsorbed species, and vapours, as described in Chapter 7. Electron diffraction should not, however, be confused with electron microscopy, in which the ability of electromagnetic lenses to focus the beams is directly exploited for the reconstruction of magnified images. Image reconstruction in diffraction methods is a highly indirect procedure owing to the loss of phase information; this is explained in Chapter 2.

Wave–particle duality holds for all material bodies, although all but the lightest particles show inconveniently small associated wavelengths. In practice only electrons and neutrons are used: hydrogen atoms and protons may be diffracted but offer no experimental advantages. Neutrons are very small and uncharged; they therefore have considerable penetrating power; they can also be produced with suitable velocities at adequate flux densities. Even though a nuclear reactor is almost always required for their production, their diffraction patterns complement those obtained with X-rays and indeed supplement them in the study of magnetic solids. Neutron diffraction is considered in Chapter 6.

2. Single crystals and X-rays

X-ray crystallography

LAUE demonstrated the synthesis of electromagnetic radiation and classical crystallography; W. L. Bragg (1913) performed the first structural analysis by single-crystal X-ray diffraction, proving that zinc blende had a face-centred cubic (f.c.c.) structure. His paper contains the germs of our present practical approach to X-ray crystallography. From it we may isolate *Bragg's Law* which is conceived in terms of reflecting planes, in contrast to Laue's conception of the three-dimensional diffraction grating. Consider Fig. 2.1, in which parallel beams of X-rays are reflected from a pair of parallel planes, just as light rays would be from a pair of mirrors. There is a clear analogy with Fig. 1.2, except that $\phi = \theta$, as required for simple reflection; Bragg's Law therefore takes the from of eqn (2.1). The value of such an approach can only be realized however, if

$$n\lambda = 2d \sin \theta \tag{2.1}$$

the three-dimensional array of scattering points can be described in terms of reflecting planes. This at once involves the true nature of crystals.

Single crystals

What is a crystal? Its outward nature is fairly simple: crystals of a particular element or compound tend to display a generally similar external morphology

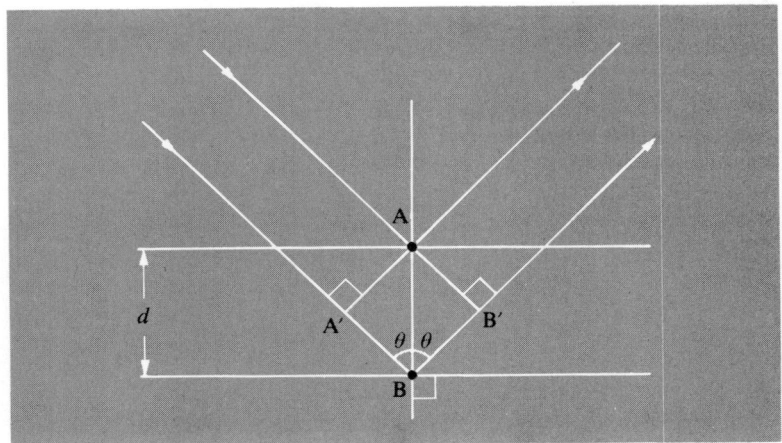

FIG. 2.1. Reflection from Bragg planes.

—that is, the same combination of plane faces with constant relative sizes, shapes, and orientations. Crystals are abundant in nature and geometrical crystallography with quantitative measurements dates from Stensen's postulate of the constancy of interfacial angles in quartz crystals in 1669. That external morphology was tied to chemical composition thus was evident at an early date, although the exact relationship was not explained until the 20th century. In 1784 Haüy proposed that regular arrangements of small polyhedra led to the regular exterior features of crystals, and today we accept the existence of such polyhedra through the concept of the *unit cell*. This must itself be composed of something, presumably of atoms or molecules. Rutherford and Marsden demonstrated the compactness and density of the nucleus by the deflection of α-particles in 1913; by then, the part of individual atoms in the construction of crystals had become accepted. Bragg's discussion of 1913 could therefore centre on a choice of packing arrangements for atoms in the structure of zinc blende.

Now regular packing in any form implies the existence of long-range order within the material and only such order can make cooperative diffraction possible. The ordering can always be broken down into multiple repetitions of the same basic pattern or motif, produced by spatial translations along one or more of three axes. The three shortest translations define the edges of the unit cell. For a given array, different choices of unit cell may be made. This is illustrated for a two-dimensional array of points in Fig. 2.2. In (a) and (b) two possible *primitive* unit cells have been selected: these contain the lowest possible number of points, viz. one. In (c), a double-sized *centred* cell has been chosen; such a choice is sometimes forced by the nature of the packing. The reader may convince himself that an infinite variety of different unit cells is possible; in practice, one takes the simplest, so that (a) would be preferred over (b) because its angles are closer to 90°. This is one of a number of conventions applied in X-ray crystallography. The unit cell is defined geometrically by the lengths of its edges, the *unit-cell axes a, b*, and *c* together with the interaxial angles α (between *b* and *c*), β (between *a* and *c*), and γ

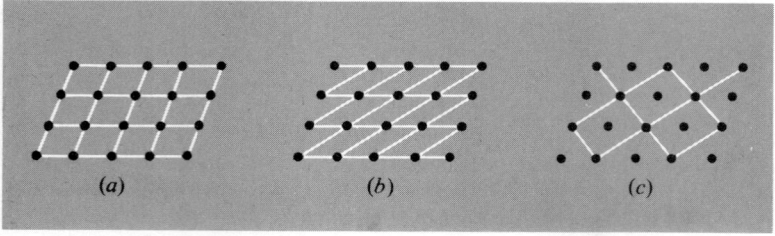

FIG. 2.2. Primitive and centred unit cells in a two-dimensional lattice.

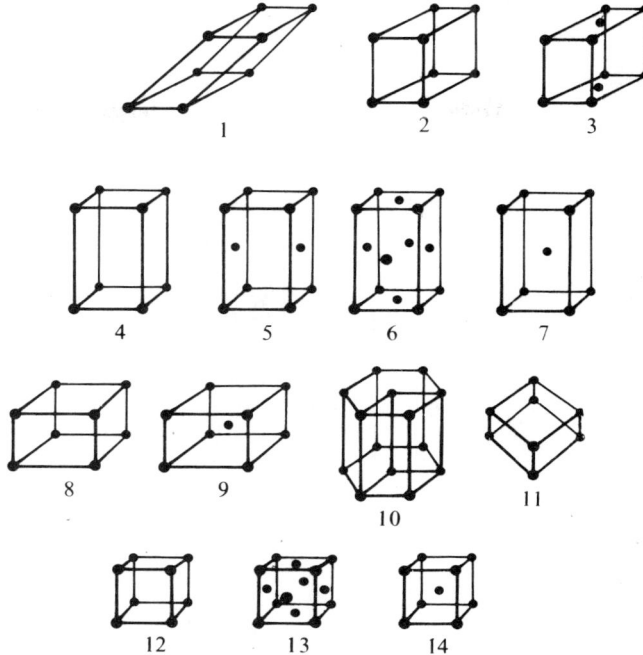

FIG. 2.3. The unit cells in the fourteen Bravais lattices.

(between *a* and *b*). Translational repetition generates the *crystal axes*. Note that the points in Fig. 2.2 have deliberately not been identified with any chemical or physical entity.

The lattice of Fig. 2.2 is only one of a number of possible varieties. Similarly, there are a number of different possible lattices in three dimensions—the fourteen *Bravais lattices*, to be exact. They are shown in Fig. 2.3. Each Bravais lattice must belong to one of seven coordinate systems: the distribution is given in Table 2.1. Inspection of Figs. 2.2 and 2.3 and of Table 2.1 should suggest that the various lattices possess different degrees of *symmetry*. Now symmetry considerations have proved very valuable in chemistry and indispensable in crystallography; we shall be concerned with the results, rather than the derivations, although it is worth noting that the latter were obtained on strictly theoretical grounds. For our purposes, a brief consideration of *point groups*, as applied to molecules and lattices, and of *space groups*, must suffice. Fig. 2.4 shows the *symmetry elements* in the ethylene molecule, which belongs to the point group D_{2h}. The 2 refers to the highest proper rotation axis, which lies along the C=C bond: a rotation of $2\pi/2$ radians about this

TABLE 2.1

The fourteen Bravais lattices and conventional coordinates

Coordinate system	Axes and angles	Centering
1. Triclinic	$a \neq b \neq c, \alpha \neq \beta \neq \gamma$	Simple
2. Monoclinic	$a \neq b \neq c, \alpha = \gamma = 90° \neq \beta$	Simple
3. Monoclinic	$a \neq b \neq c, \alpha = \gamma = 90° \neq \beta$	Side-centred
4. Orthorhombic	$a \neq b \neq c, \alpha = \beta = \gamma = 90°$	Simple
5. Orthorhombic	$a \neq b \neq c, \alpha = \beta = \gamma = 90°$	End-centred
6. Orthorhombic	$a \neq b \neq c, \alpha = \beta = \gamma = 90°$	Face-centred
7. Orthorhombic	$a \neq b \neq c, \alpha = \beta = \gamma = 90°$	Body-centred
8. Tetragonal	$a = b \neq c, \alpha = \beta = \gamma = 90°$	Simple
9. Tetragonal	$a = b \neq c, \alpha = \beta = \gamma = 90°$	Body-centred
10. Rhombohedral	$a = b = c, \alpha = \beta = \gamma \neq 90°$	Simple
11. Hexagonal	$a = b \neq c, \alpha = \beta = 90°, \gamma = 120°$	Simple
12. Cubic	$a = b = c, \alpha = \beta = \gamma = 90°$	Simple
13. Cubic	$a = b = c, \alpha = \beta = \gamma = 90°$	Face-centred
14. Cubic	$a = b = c, \alpha = \beta = \gamma = 90°$	Body-centred

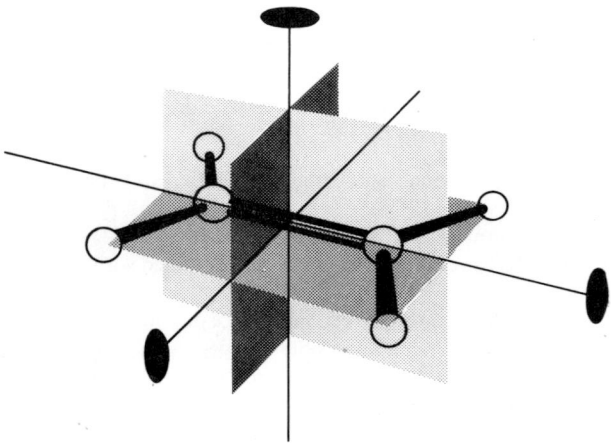

FIG. 2.4. The symmetry elements in the ethylene molecule, three two-fold axes and three mirror planes.

axis puts the molecule in a position indistinguishable from its original orientation. In general, an n-fold proper rotation axis refers to a rotation of $2\pi/n$ radians. The D refers to the existence of n further two-fold axes in a plane normal to the principal axis and the h to a mirror plane normal to the principal axis. The two-fold axes are represented by the symbols shaped like American footballs and the mirror planes are shaded. Note that this combination of symmetry elements automatically gives rise to two further mirror planes containing the principal axis. The operations of rotation and reflection are called *symmetry operations*. The molecule also contains a centre of symmetry. Symmetry elements and operations, as well as point-group symbols, are fully discussed by Cotton (1966). The molecule does not change its position in space under symmetry operations. A lattice also has point-group symmetry, provided it extends indefinitely in all directions. That of Fig. 2.2 evidently contains a centre of symmetry at each of the points : that is, if we consider any point as the origin, any feature at coordinate position (x, y) is matched by one at $(\bar{x}, \bar{y})$. For lattices, however, there is a limit on the kinds of symmetry elements that may exist, notably proper axes. In fact, there are only thirty-two possible combinations of symmetry elements about a fixed point, corresponding to the thirty-two *crystallographic point groups*. When these elements are combined with translations of the unit cell, the 230 space groups are produced. These contain translational symmetry elements which, in contrast to the point-group elements, can be detected by features of the diffraction patterns of crystals, as explained in Chapter 3. The point groups and space groups of crystallography are tabulated in the *International tables for X-ray crystallography* (1965), vol. I ; this volume is worth a glance, if only as something of a typographical masterpiece.

Diffraction from lattices

For the Bragg equation to serve the analysis of diffraction patterns, the Bragg planes in crystals must be identified. Look again at Fig. 2.2 : using lines as two-dimensional analogues of planes, any number of them can be drawn through the lattice. Selecting any one we realize that it must cut through some unit cell; its orientation can be defined with respect to the unit-cell axes. Translation of the unit cell generates the whole lattice ; translation of the line that cuts it generates a *family* of lines, with equidistance between nearest neighbours. These families of hypothetical planes, given physical substance, should produce Bragg reflections. To find the directions of the incident and reflected beams of Fig. 1.5, we need to specify the direction and magnitude of AB in that Figure. To do this with respect to the cell axes, imagine that one plane of the family traverses the origin of a unit cell, as in Fig. 2.5. It is convenient to specify AB in terms of the intercepts of the next neighbour in the family on the cell axes. If for the three-dimensional case the fractional intercepts are X/a, Y/b, and Z/c, or (x, y, z), they may also be quoted as reciprocal fractional intercepts $(1/x, 1/y, 1/z)$. Clearing the fractions gives the *Miller indices* of the plane, (h, k, l). If we know a, b, c, α, β, and γ, the indices $(h\,k\,l)$ completely specify the orientation and separation of the planes in that particular family. The factor n in Bragg's Law is absorbed into these indices. For any given value of λ, Bragg's Law sets a limit to the number of families of

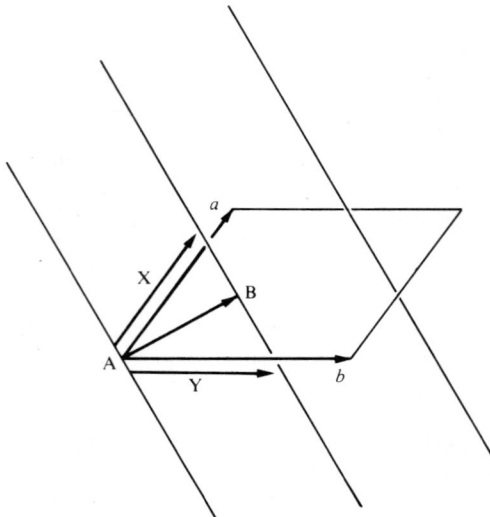

FIG. 2.5. Intercepts on the unit-cell axes.

interest: θ cannot exceed $\pi/2$ radians, so that |AB| cannot exceed $\lambda/2$. This sets a limit of *resolution* of scattering planes or centres within the lattice for any particular diffraction experiment.

Crystallographers frequently talk about *reciprocal space* and *reciprocal lattices*. These are not abstruse concepts of relativistic physics but references to a convenient geometrical construction. If the lines AB for all the families of planes are plotted out from the origin, their far ends are found to form a mini-lattice within the confines of the unit cell. If a set of lines designated as AB* and having the same orientation as AB but a length λ/AB are plotted, their far ends form an array of points called the reciprocal lattice. The real, or *direct* lattice and the reciprocal lattice have a fixed mutual orientation. The reciprocal lattice is used in the Ewald construction of Fig. 2.6. Here, OA is the radius of a circle of unit radius, while AB* is one of the lines defined above, with A at the origin of the reciprocal lattice. Since CA has a length of 2 units, $\sin(ACB) = AB^*/2 = \lambda/2AB$. The angle AOB can thus be identified with the Bragg angle 2θ, while CO and OB* coincide with the incident and re-flected beam directions. If, therefore, a reciprocal-lattice point B* touches the surface of the circle, the Bragg condition is met and a reflection is developed. The circle is called the *circle of reflection*; extending the argument to three dimensions, it becomes the *sphere of reflection*. The condition can only be met for AB* $\leqslant$ AC, i.e. AB* $\leqslant$ 2. There is thus a *limiting sphere* of radius 2 which contains all the reciprocal-lattice points accessible to the experimenter

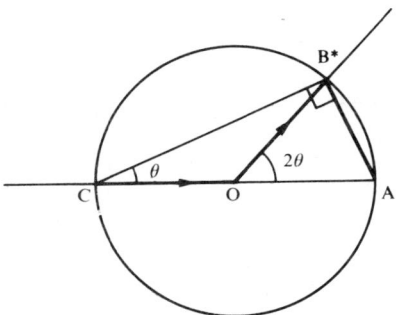

FIG. 2.6. The Ewald construction

with the given value of λ and corresponding to the limit of resolution in the direct lattice.

Atoms in lattices

The lattice has so far had an incorporeal existence and has been no more than the skeleton defining the spatial repetitions of an unspecified motif. In reality the motif consists of an assembly of atoms or ions, forming part or the whole of a molecule or formula unit. It is called the *asymmetric unit*. Consider sodium chloride: this actually contains two interpenetrating face-centred cubic (f.c.c.) lattices, one of Na^+ and one of Cl^- ions, as illustrated in Fig. 2.7. Both lattices have exactly the same unit-cell dimensions and we can define them to have an ion at each corner and face-centre of the cube. Either lattice would *independently* give rise to the same set of Bragg reflections, with incident and diffracted beams in the same orientation with respect to the cell axes for a

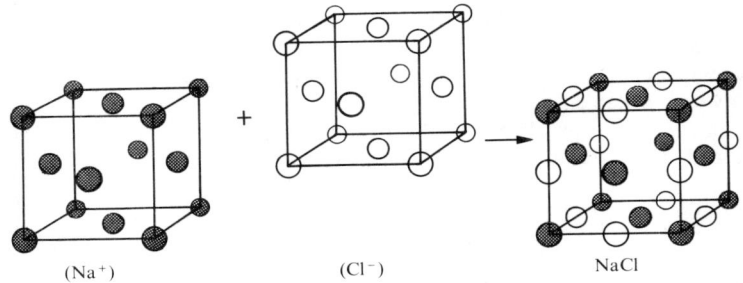

(Na⁺)   (Cl⁻)   NaCl

FIG. 2.7. The sodium-chloride lattice.

given index triplet (*hkl*). There is, however, one important difference between the two : the diffracted beams are not in phase. The phase difference is controlled by the separation between the origins of the lattices. For an asymmetric unit consisting of many atoms, a similar lattice can be constructed with its origin centred on each atom in turn. For any particular Bragg reflection, a summation of contributions from the different lattices can be carried out to give the *structure factor* F_{hkl}. The expression for the structure factor is a function of a number of variables, as shown in eqn (2.3). The symbol $\sum_j$ indicates the summation over all the atoms in the unit cell, rather than over those in the

$$F_{hkl} = \sum_j \mathfrak{F}(f_j, B_j, \theta, \lambda, h, k, l, x_j, y_j, z_j) \qquad (2.3)$$

asymmetric unit, in order to allow for the space-group symmetry. f_j is the scattering amplitude of the *j*th atom, or *form factor*. (x_j, y_j, z_j) are its fractional coordinates with respect to the unit-cell axes and B_j a parameter that measures its amplitude of thermal vibration.

In using the form factor, f_j, we have tacitly assumed that the atom is the effective scattering centre, which acts as a small point. This is not exactly the case, as the actual scattering occurs at the electrons surrounding the nucleus, which are forced into oscillation by the electromagnetic field of the incident X-ray. The scattering centre is therefore better described as a spheroid, whose scattering power diminishes with increasing distance from the nucleus. As this electron distribution has finite dimensions, comparable to normal X-ray wavelengths, an interference effect arises between the different parts of it. This effect will be a function of θ and λ. Luckily, the chemical environment of an atom or ion, although affecting its valence shell, has little influence on the greater part of its electron distribution, the core electrons. We can therefore use the same distribution for any one atom or ion under virtually all circumstances. Consequently, eqn (2.3) can be used to construct a mathematical model of the scattering effects of a known crystal structure. The value of f_j is usually drawn from a set of tables calculated on the basis of semi-empirical data—from the self-consistent field (SCF) model of the atom, for example. If, therefore, the unit-cell parameters, the space group, the nature, position, and vibration amplitude of all the atoms are known, the orientations and amplitudes of all the Bragg reflections accessible to radiation of a given wavelength may be calculated.

Eqn (2.3) is all very fine, provided one knows the crystal structure. What is needed, of course, is an equation such as eqn (2.4), which is essentially the

$$(x_j, y_j, z_j, B_j) = \mathfrak{f}(F_{hkl}) \qquad (2.4)$$

reverse of eqn (2.3). There is no such direct relationship in terms of atomic positions and vibration amplitudes. If, however, the unit-cell contents are described in terms of electron density, without reference to atoms, it is possible to derive an expression for this density, $\rho(x, y, z)$ at any point (x, y, z) in the unit cell. This is eqn (2.5), in which the summation is over all the reflections

$$\rho(x, y, z) = \mathfrak{f}(\sum F_{hkl}, h, k, l, x, y, z, V) \qquad (2.5)$$

and V is the unit cell volume. All we need to do, then, is to measure all the values of F_{hkl} and calculate enough local values of ρ to draw up a map of electron density in the unit cell. This map will then show the atomic positions and the structure will be solved. There is, needless to say, a catch, which is explained in the next section. For the more mathematically inclined reader, the forms of eqns (2.3) and (2.5) are given in Appendix II.

The phase problem

The catch is that F_{hkl} is a quantity corresponding to a sum of oscillatory wave motions. Those who have studied waves in sound or alternating currents will recall that they have both amplitudes and phases. X-ray detectors unfortunately measure cumulative intensities over time, I_{hkl}: the phases cannot be detected. All that can directly be found is the modulus of the structure factor $|F_{hkl}|$. The central problem of X-ray crystallography, the *phase problem*, is that of retrieving these phases from the set of moduli. This is generally done by some sort of 'bootstrap' operation. There are different varieties of operation which are often used together or in turn, as the crystallographer becomes increasingly desperate; every piece of contributory evidence, chemical or otherwise, is a help. Four main approaches are in common use.

Trial and error

No two molecules can be in the same position at once, nor two atoms. If the shape of the molecules in the unit cell is already roughly known, and particularly if they are complex and elongated, enlightened guesses may be made as to the correct positions and orientations. The method was valuable, in the early years, for highly regular salts and minerals, whose atomic positions were determined by simple considerations of packing.

Patterson maps

These maps are calculated from an equation similar to eqn (2.5) and are named after its discoverer. The difference is that F_{hkl} is replaced by F_{hkl}^2, for which the phase no longer matters. The maps represent the relationships between all the different atoms in the unit cell—or, more exactly, all the elements of electron density. Suppose, for example, that there are two atoms at (x_1, y_1, z_1) and (x_2, y_2, z_2). The Patterson map shows other atoms surrounding the first taken as the origin: a peak appears at $(x_2 - x_1, y_2 - y_1, z_2 - z_1)$. It also does the same for the second atom, so that another peak appears at $(x_1 - x_2, y_1 - y_2, z_1 - z_2)$. As the number of atoms increases, the pattern of peaks centred on the origin rapidly becomes more complex—in principle, these should be $n(n-1)$ peaks for n atoms. The problem is to deconvolute this pattern and work back to the original arrangement of atoms. This can be almost impossible for structures containing large numbers of atoms of relatively even scattering power. Fortunately, symmetry elements in the unit cell lead to some simplifications and the form of the Patterson relationship causes peak heights to be proportional to the product of the scattering powers of the two atoms involved. Since this scattering power is proportional to atomic number, the interactions for a compound such as ferrocene ($C_{10}H_{10}Fe$) will have roughly these relative strengths:

$$Fe-Fe = 26 \times 26 = 676$$
$$Fe-C = 26 \times 6 = 156$$
$$C-C = 6 \times 6 = 36$$

Interactions involving hydrogen are all weaker than these. The Fe–Fe relationships therefore stand out very strongly and the iron-atom position can readily be discovered. This 'heavy-atom' method has found wide use as an entry to phasing reflection data, although it becomes less and less easy to apply when several heavy atoms are present or when accidental near-symmetries exist in the unit cell.

Direct methods

Theoretical analyses show that certain probability relationships exist between the signs or phases of reflections having certain combinations of indices. The theory is too complex to discuss here but it has been put into practice successfully. The first applications were to centrosymmetric structures —i.e. those containing a centre of symmetry—because the structure factors from these have phases of 0 or π radians, corresponding to simple plus and minus signs. More recently, non-centrosymmetric structures have been solved by direct methods, or statistical phasing, as they are also called. Organic compounds may usefully be tackled in this way : they often crystallize in non-centrosymmetric structures, and they contain no outstandingly strong scatterer, unless one is deliberately introduced. The even distribution of scattering power throughout the unit cell is in fact a basic assumption of the theory of direct methods.

Isomorphous replacement

Although it may be possible to introduce a heavy atom into a compound by substitution or other means, this may not suffice as an entry to phasing the complete set of reflection data (see p. 17). If, however, a heavy atom can be introduced without disturbing the chemical or crystal structure, thereby producing isomorphous although chemically different forms, two sets of data can be collected and compared. By observing which reflections show a gain or loss in intensity, one may deduce the relative phases of the contributions from the heavy atom and from the rest of the structure. The original and classic application was made by Robertson (1936) in the study of phthalocyanines. The method has since proved very useful indeed for large structures such as those of proteins.

Extension of phasing

The methods quoted above provide only a partial solution of the structure. Trial-and-error positioning is at best approximate; Patterson methods usually reveal only the heavier atoms; statistical phasing and isomorphous replacement provide phase information for the strongest reflections of the data set only. Fortunately, the very fact of the concentration of the electron density around the atomic centres makes extension of the phasing possible. The heavier concentrations effectively dominate the phasing of all reflections; consequently, the heavy-atom contributions can be used as a preliminary guide to the phases of all reflections. Furthermore, approximately correct sets of atomic coordinates provide adequate phasing models for the structure to be solved by a series of successive approximations.

The usual procedure is to use whatever atomic positions and atomic scattering amplitudes are known in a *structure-factor calculation*, using eqn (2.3). The phases calculated from this are then applied to the observed (and hitherto unphased) amplitudes, which are in turn used to produce a map of electron density, by means of eqn (2.5). Such a map is called a *Fourier map* after the mathematical approximations used in eqn (2.5) and first developed by Fourier. It should show peaks of positive sign at the positions of greatest electron density; the heights of these peaks are a measure of the atomic numbers of the atoms at these positions. If all goes well, the Fourier map will not only show the atoms already introduced but also peaks corresponding to further undiscovered atoms. If direct methods have been used, one begins with a map rather than a structure-factor calculation. As additional atoms are located, they must be assigned sensible chemical identities, while peaks in obviously ridiculous locations can be temporarily ignored. The new set of atomic identities and coordinates is then entered into a new structure-factor calculation and the whole process is cyclically repeated until all the atoms have been identified and placed. The process of searching Fourier maps can profitably be continued until the structure is completely solved and refined— it is by no means unusual, for example, for an unsuspected molecule of solvent of crystallization to reveal itself at a very late stage. Chemical good sense has to be used throughout and it may be necessary to try several different models. As an overall guide to the success of the model, the '*R*-factor' or discrepancy index is much used. This is obtained from the structure-factor calculation, in which it is convenient to list the observed and calculated structure factors side by side; the algebraic sum of the discrepancies for individual structure factors, divided by the sum of the structure factors, is the R-factor. It should normally decrease steadily as the model improves in realism.

Refinement

All atoms have now been identified, approximately positioned, and given rough amplitudes of thermal vibration. Note that pseudo-solutions are quite possible and can cause a good deal of trouble and frustration. The last stage is the refinement of the model to the point of which the discrepancies between observed and calculated structure factors, $F(\text{obs})$ and $F(\text{calc})$, are minimized. One tool for this purpose is the *difference Fourier map*, based on a special structure factor defined as $F = F(\text{obs}) - F(\text{calc})$. This map shows positive peaks where electron density exists but has not been specified in the structure factor calculation, and negative peaks, or 'holes', where unwarranted scattering power has been built into the model. Thus a slightly misplaced atom produces a peak on one side and a hole on the other. One adjusts coordinates and thermal vibration parameters until a difference Fourier map with as few 'features' (peaks or holes) as possible is produced.

The more powerful alternative is purely arithmetical: this is the *least-squares method*, by which a direct minimization of discrepancies in structure factors is obtained by altering the structural parameters. This procedure relies on the overdetermination inherent in single-crystal work: a good data set provides fifteen to thirty quantitative observations for every parameter in the model. This is why crystallographers can provide such detailed and precise information on molecular dimensions. Conversely, one should beware of pursuing a low R-factor by the introduction of parameters not justified by the quality of the primary data. Fortunately the referees who scrutinize crystallographic papers before publication are usually eager to pounce on such technical failings in their colleagues or rivals.

Conclusion

Solving structures may seem to be a pleasantly logical process, to judge from the synopsis just given. The practical bugbears have been relegated to Chapter 3. Before we proceed to them, it is worth stressing that solving crystal structures is by no means universally simple. Modern aids in data collection and processing have greatly alleviated the former tedium but a substantial degree of skill and experience is still required. X-ray crystallography today is the child of the match between old skills and intuitions and modern technical developments, notably the high-speed digital computer.

PROBLEMS

2.1. The largest faces of a crystal do not usually correspond to the direction of fastest growth. Why?

2.2. The HCl molecule has an ∞-fold proper axis of rotation and the IF_7 molecule (a pentagonal bipyramid) a fivefold axis. Why can these symmetry elements not be reproduced in the crystal structures of solid HCl and solid IF_7?

2.3. Show that Bragg's Law and the three Laue equations are equivalent.

2.4. Space group $P2_1/c$ has the following general positions related by symmetry: (x, y, z), $(\bar{x}, \bar{y}, \bar{z})$, $(\bar{x}, \frac{1}{2} + y, \frac{1}{2} - z)$, $(x, \frac{1}{2} - y, \frac{1}{2} + z)$. Where would you expect to find peaks in the Patterson map for interactions between symmetry-related atoms? What relative heights would they have?

3. Practical structure determination

The production of X-rays

WHEN electrons are accelerated through a potential difference of 20 to 50 kV to fall onto a metal anode, the anode emits a continuous spectrum of 'white' X-radiation. Superimposed on this are sharp and intense X-ray peaks whose frequencies are characteristic of the metal employed. These peaks arise from the forcible ejection of a K-shell electron from the atom under bombardment and the collapse of an L-shell or M-shell electron into its place, with emission of an X-ray quantum. As the nuclear charge increases, so does the energy of the quantum and therefore the frequency of the peak. The Laue photographs were originally taken with white radiation but it is now more usual to use the characteristic lines, usually the K_α doublet (K ← L). Modern X-ray tubes use a water-cooled metal target which, for diffraction purposes, is viewed at an acute angle to its surface in order to minimize its apparent area. The commonest target materials are copper and molybdenum, providing K_α lines at 154·12 and 71·07 pm, respectively. A typical X-ray generator is shown in outline in Fig. 3.1. The K_α line is selected either by inserting a thin metal-foil filter in the port through which the X-rays emerge, or by constructing a monochromator that uses a strong Bragg reflection from a suitable crystal.

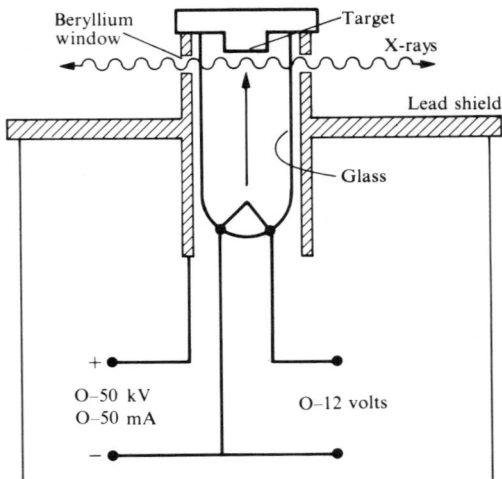

FIG. 3.1. Outline of an X-ray generator: X-rays leave the evacuated glass tube through thin beryllium windows; the target is water-cooled through the lead 'tower'.

Since X-rays cannot be focussed, a thin and approximately parallel pencil beam of uniform intensity cross-section is selected by means of a collimator consisting of a narrow metal tube.

Three principal considerations dictate the choice of target. The shorter the wavelength, the greater the region of reciprocal space accessible to diffraction. Conversely, shorter wavelengths lead to lower Bragg angles, which may lead to inconveniently small separation of reflections from the larger unit cells. Finally, shorter-wavelength X-rays are more penetrating and suffer less absorption in the specimen. The absorption coefficient for a given wavelength is dictated by the proximity of the atomic number of the absorber to that of the target. When the atomic number of the absorber is lower than that of the target, the absorption increases rapidly as the characteristics of the two become increasingly similar, until an 'absorption edge' is reached for the element one below that of the target. For an equal or higher atomic number, absorption drops dramatically. Thus nickel screens out most of the K_β line (K $\leftarrow$ M) of copper and makes a good filter, while zirconium or niobium are used with molybdenum targets. Correspondingly, copper K_α radiation is best avoided with specimens containing iron, cobalt, or nickel, as much of its energy will be dissipated in unwanted fluorescence.

Determining the unit cell and space group

We discussed above how incident and diffracted beam directions are related to the unit-cell axes, using the reciprocal-lattice concept and the Ewald construction. The first step in any structure determination is therefore the identification and measurement of the unit-cell parameters, which may or may not be traceable through the external morphology of the crystal. Techniques vary but a common starting point is the taking of an *oscillation photograph*; the principle is explained in Fig. 3.2. A suitable crystal is chosen, typically with dimensions of the order of $0.1 \times 0.1 \times 0.1$ mm, so that it may be totally immersed in the primary X-ray beam. It is stuck to a thin glass fibre with glue or wax or, if it is unstable to air, mounted inside a very thin-walled capillary tube. The fibre or tube is itself stuck to a small metal pin, mounted in a *goniometer head*, which is a device having two arcs, mutually at 90°, which allow minor angular adjustment of crystal orientation with respect to the rotating shaft which carries the goniometer head. The primary beam emerging from the collimator bathes the crystal and is then caught by a lead backstop. A cylindrical film encircles the shaft axis, or mounting axis, and intercepts and records any diffracted beams. The shaft is made to perform small rotatory oscillations, of say $\pm 10°$, so that a limited number of reciprocal lattice points cut the sphere of reflection. After an exposure of 20 to 30 minutes with normal generators and X-ray films, the film is unrolled and developed. If the crystal has been mounted so that a principal unit-cell axis coincides with the shaft axis, a photograph which looks like Fig. 3.3(a) results. *Layer lines* of spots are

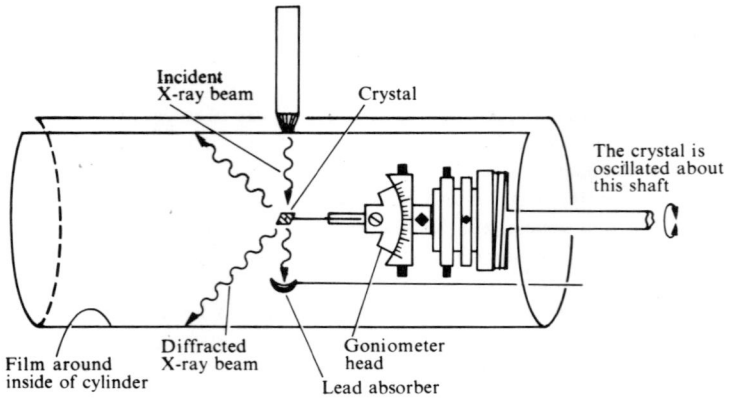

FIG. 3.2. Taking an oscillation photograph of a single crystal.

observed; these correspond to families of reciprocal lattice points lying in planes normal to the mounting axis. For a b-mounting, or mounting along the b-axis of the unit cell, the central line contains spots with the indices $h0l$; the pair of lines on either side correspond to $h\pm1l$. The reader is left to convince himself of this by considering the relationship between the direct (real) and reciprocal lattices. Should the alignment of the crystal be imperfect, as it usually is, a photograph that looks like Fig. 3.3(b) results. The angular misalignment can be measured from it and two such photographs taken at shaft positions 90° apart can be used to determine the necessary corrections to be applied to the goniometer arcs. The corrections may be refined by stages, until an accuracy of $\pm5'$ of arc is achieved. Since unit-cell axes are often reflected

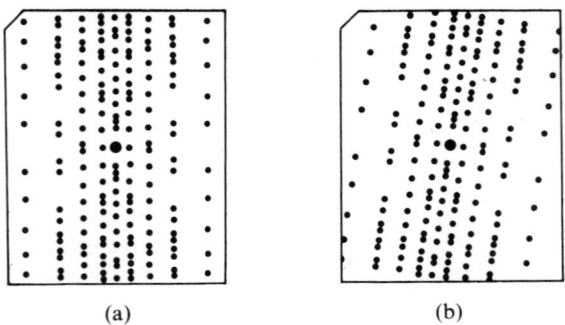

FIG. 3.3. Oscillation photographs, (a) correctly aligned, (b) misaligned.

in external morphology, one can sometimes pick a suitable mounting by examining the crystal under a microscope. If not, the oscillation photograph provides an acquisition range of up to 20° of misalignment, so that a direct cell axis can usually be found quite quickly.

Once the alignment is correct, a *rotation photograph* is taken. For this, full 360° rotation of the shaft is allowed. The photograph is similar to that of Fig. 3.3(a), except that the full families of $h0l$ etc. in the layer lines now appear, subject to the overriding limit of resolution. In principle, therefore, the rotation photograph contains all the information required to solve the structure. In practice, there is no simple way of assigning the h and l indices to the various reflections, since there is no way of associating any one of them with the angular position of the shaft or the a and c axes. There can also be superimposition of reflections with different h and l but the same Bragg angle. Consequently, the rotation photograph is generally used only to determine the length of the mounting cell axis, from the spacing of the layer lines, and the Laue class, or lattice symmetry, of the crystal. The limitations are ingeniously overcome in the Weissenberg camera. Here, the shaft rotation is coupled to a transverse motion of the film, parallel to the shaft; a slit-metal screen ensures that only one layer line at a time is examined. Exposures of about 48 hours are normally used, so that even weak reflections become visible on the developed film. In effect the Weissenberg photograph decomposes the h–l ambiguity into a transverse coordinate on the film. The vertical coordinate continues to

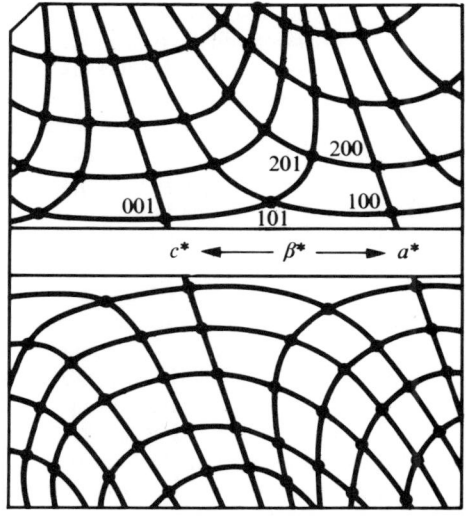

FIG. 3.4. A Weissenberg photograph.

represent the Bragg angle. Fig. 3.4 represents the $h0l$ Weissenberg photograph of a monoclinic crystal, mounted along its b axis. It is actually a distorted view of the $h0l$ plane in the reciprocal lattice. The pattern of festoons of spots is distinctive and universal; it allows recognition of principal axes and interaxial angles. Each festoon represents a line in the reciprocal lattice, as the partial indexing in Fig. 3.4 shows. The axes a^* and c^* have been lined in (the asterisk is conventionally used to denote reciprocal-lattice quantities). Knowing the wavelength of the radiation, the diameter of the camera, and the ratio of the shaft-to-camera gearing, we can obtain the values of a^*, c^*, and β^* from this photograph. For a crystal of monoclinic or higher symmetry the rotation and Weissenberg photographs are enough to determine the unit-cell parameters. By isolating the appropriate layer lines the $h1l$, $h2l$ etc. photographs are obtained. The process can be repeated for mountings about the other axes but it is often more convenient to transfer the goniometer head and crystal undisturbed to a *precession camera*.

This is a mechanically more complex device which takes photographs of reciprocal lattice planes not normal to the mounting axis. Photographs of the planes $0kl$, $1kl$ etc. and $hk0$, $hk1$ etc. can thus be obtained without change of mounting. Some crystallographers in fact prefer to start their examination of a crystal with a precession camera. When the set of Weissenberg and precession photographs is examined, the complete dimensions and symmetry of the lattice can be obtained. The precession photographs have the additional advantage of being undistorted views of the reciprocal lattice planes. From the set of photographs, the pattern of *systematic absences*—if any—can be determined. These absences are symptomatic of the presence of space-group symmetry elements. They may allow positive identification of the space group; more commonly, however, they merely narrow the possibilities to a group of space groups. In particular, they provide no way of differentiating between centrosymmetric and non-centrosymmetric space groups because the diffraction process always adds a centre of symmetry. Bragg angles are often measured for a number of reflections and the cell parameters are adjusted by least-squares fitting to match these measurements as accurately as possible.

Intensity data

The Braggs originally used electrometers to measure the ionizing power and hence the intensity of diffracted beams. The introduction of rotation and Weissenberg photographs caused this laborious method to be superseded. X-ray films with normal silver halide emulsions fortunately have an almost linear response to cumulative X-ray intensity over a wide range of flux and exposure. Each spot is therefore an accurate and linear record of integrated intensity for a particular reflection over the exposure time of the film. The problem is to read off this intensity: traditionally, this was done by visual comparison with a set of graded standard exposures on a strip of film, taking

advantage of the natural integrating properties of the human eye. In essence
the eye measures the degree of blackening; complications therefore arise with
respect to the extent and shape of the spots. Nevertheless experience, perse-
verance, and some eye-strain can give a surprisingly good accuracy of some
± 5 per cent in intensity. Data collection by film methods has definite advantages: apart from low equipment costs, all reflections are collected more or
less simultaneously, which can be a distinct advantage over a sequential
procedure if the crystal is unstable. In order to retain these advantages,
integrating Weissenberg cameras have been designed; these 'square off' the
reflections by moving the film through small intervals away from its normal
gearing to the traversing mechanism between passes. Some effort has also been
put into developing coordinate arrays to replace the film; such arrays consist
of semi-conductor detectors or imaging systems analogous to TV camera
tubes. Little success seems to have been had so far and the imaging systems
may have to return to the rotation-photograph geometry, using conventional
flat-screen tubes and special provisions for indexing. Automatic scanning of
X-ray films, using the flying-spot microdensitometer technique allied to
computerized read-out, may be more promising.

Although these considerations may presage a return to film methods of
collecting data, the major trend since 1945 has been towards the use of
diffractometers. These devices are a mechanized version of the original Bragg
method and consist of the usual X-ray source, a detector for the diffracted
beam, and means of orientating sources, crystal, and detector for the recording
of each Bragg reflection in turn. Two geometrical constructions are commonly
used. The system of Fig. 3.5(a) essentially mimics the Weissenberg or rotation
photograph, and is often called the Weissenberg geometry. The crystal is

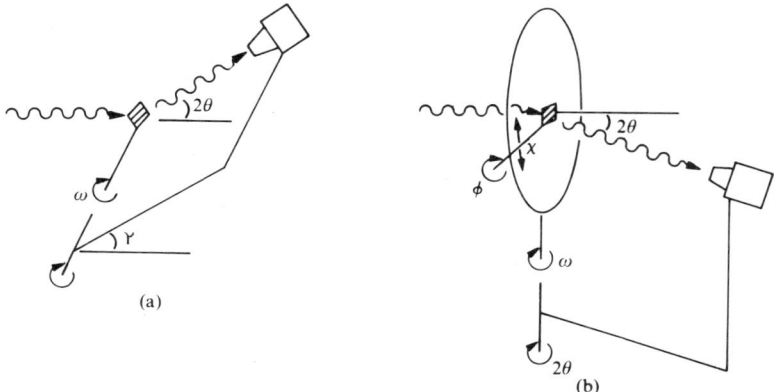

Fig. 3.5. Diffractometer geometries: (a) two-circle (Weissenberg), (b) four-circle.

usually mounted along a major direct-cell axis and can be rotated through 360° about it. This axis and the incident beam define a basal plane; the angles within the plane between the incident beam, the crystal mounting axis, and the crystal–detector axis (diffracted beam) are preset for the collection of each level of data, reproducing the inclination settings of the Weissenberg camera. By elevating the detector to an angle Υ out of the basal plane and rotating the crystal about the mounting axis to the appropriate angle (ω), the various Bragg reflections in the level can be recorded by the detector. The vertical sweep of the detector aperture reproduces the slit screen of the Weissenberg camera. Reflections are recorded one at a time, so that 2θ coincidences are decomposed by time, rather than by lateral motions of a camera body. As two angles must be set for each reflection, this is commonly called a 'two-circle' diffractometer.

In the four-circle geometry of Fig. 3.5(b), incident beam, crystal, diffracted beam, and detector all lie in the basal plane; the Bragg angle 2θ is set by moving the detector about the vertical $\omega/2\theta$ axis. The desired family of planes is then brought into the reflecting orientation by a combination of three adjustments: the goniometer head travels around the vertically-mounted χ-circle, to an angle χ; the χ-circle rotates about its mounting axis to an angle ω; the goniometer head rotates about the mounting axis to an angle ϕ. All four settings are continuously adjustable, although the χ-circle is sometimes limited to a quarter circle. The need for good alignment in such a complicated mechanism calls for very high standards of construction and machining and adds substantially to the cost of the diffractometer. The principal benefit is one of flexibility; the crystal can be mounted along any arbitrary axis and the experimenter can control the angle at which the sphere of reflection is cut as the crystal moves into the Bragg condition. The four-circle geometry therefore allows the same angle of attack to be maintained for all reflections, which is impossible on two-circle machines; the data recorded is correspondingly more accurate.

The importance of the angle of attack arises from the nature of the detector: this is normally a scintillation or gas-proportional counter, often capable of detecting single X-ray quanta, feeding appropriate discriminating, amplifying, and counting circuits. In contrast to a film, it measures continuous X-ray flux, so that integrated intensities must be obtained from timed exposures; furthermore, the finite mosaicity of the crystal generally requires an angular scan through the Bragg condition, with a length controlled by the angle of attack to the sphere of reflection.

The background random scattering close to but not at the Bragg peak is normally measured and its intensity subtracted from that of the peak. The design and mode of operation of diffractometers are discussed in detail by Arndt and Willis (1966). Various degrees of automation are possible, ranging from hand-setting of the various angles through digital control by punched cards or tapes to full on-line computer control. Many of the positioning devices and programs are akin to those used in the numerical control of machine tools. In all cases the indices of the Bragg reflections, the unit-cell dimensions, and the X-ray wavelength are required as input to the calculations, whether conducted on-line or separately.

Data reduction

By this stage a considerable amount of calculation has already been under-taken—a medium-size structure is likely to involve the collection of two thousand reflection intensities for which diffractometer setting angles may have to be produced. X-ray crystallographers therefore make extensive use of digital computers, which become increasingly valuable as the structural analysis proceeds. This is a comparatively recent development but one which has, for all practical purposes, become indispensable. The pre-computer methods used were very ingenious and remain instructive but, for reasons of space, we shall confine ourselves to considering current practice. However one proceeds, the first step is data reduction, or the conversion of the raw intensities, the set of $I(hkl)$, into F(obs) on a common scale. Since these two quantities are not simply proportional, a good deal of computation is involved. As the reciprocal lattice is rotated in conjunction with the crystal, the points that make it up cut the sphere of reflection at different angles of attack. Ideally, the sphere is an infinitesimally thin shell, so that the Bragg condition is satisfied for an equal and infinitesimally small time. In actual fact real crystals are not perfectly ordered throughout; a better approximation is to think of them as a mosaic of small, perfect blocks slightly misaligned with respect to each other. This property, together with the finite width of the spectrum of the incident beam, which is never perfectly monochromatic, make for a finite thickness of the sphere of reflection. Diffraction therefore occurs over a small angular range. Again, interference between the different parts of the electron distribution of the atom causes a polarization of the diffracted beam. These two effects require a *Lorentz-polarization* (or LP) *correction* to be applied to the raw $I(hkl)$; this is made purely on the basis of the diffraction geometry and can therefore be exactly calculated.

Two less tractable corrections to the $I(hkl)$ remain: extinction and absorp-tion. Extinction is the weakening of the incident X-ray beam by the coherent diffraction process as its traverses the crystal. Because of the relatively low scattering power of atoms for X-rays, it can usually be ignored except in the more extreme cases. Absorption is the result of wastage of intensity through incoherent scattering; this becomes increasingly great as the atomic numbers of the atoms in the specimen rise and is especially acute near absorption edges, which are best avoided. Absorption is governed by an exponential relation-ship given in eqn (3.1), which relates the original intensity I_0 and the remaining intensity I after the beam has penetrated a distance x into the crystal. The constant μ is the *absorption coefficient*, a characteristic of the material in the crystal. Equation (3.1) is in fact the usual Lambert–Beer Law governing the

$$I = I_0 \exp(-\mu x) \qquad (3.1)$$

absorption of electromagnetic radiation. To obtain the desired correction we must consider diffraction as it occurs from each infinitesimally small element of the crystal, dV. If the path distance of the incident ray through the crystal to dV is p_i and that of the diffracted ray from dV is p_d, one can obtain a mean value of the transmission coefficient, T, by integration over the whole volume of the crystal, V. This takes the form of eqn (3.2). The problem is to find an

$$T = (1/V) \int_0^V \exp\{-\mu(p_i + p_d)\} \, dV \qquad (3.2)$$

expression for p_i and p_d in terms of the position of dV within the crystal. Unfortunately, there is no general expression for this, unless the crystal is a cylinder or a sphere. Suitable crystals, particularly mineralogical specimens, can be ground to these shapes but fragile or unstable materials hardly lend themselves to such handling. A solution is found in numerical integration, analogous to the use of Simpson's Rule for the area under a curve. The general procedure is to slice up the crystal—figuratively, not literally—in three mutually perpendicular directions, so as to divide it into microscopic brick-shaped elements. The coordinates are related to the positions of the corners of the crystal, whose exterior shape must be approximately described by bounding plane faces. p_i and p_d for each (hkl) triplet and each little brick can then be calculated; the transmission coefficient for each brick is found and T is taken as the mean for all the bricks. Six or eight slices along each co-ordinate direction are required to achieve sufficient accuracy in T, creating 216 or 512 bricks. The intersection of the incident and diffracted beam from each brick with the planes of every bounding face must be found—the correct one is that lying closest to the centre of the brick. For a cubic crystal, having six faces, this requires $2 \times 6 = 12$ calculations per brick, or 2692 or 6124 for each reflection! Clearly a superhuman task by hand and still quite lengthy even on the fastest computers. That the task is worth undertaking is strongly suggested by the data of Table 3.1. To ignore a correction of such magnitude is to throw away the painstaking efforts expended in collecting accurate sets of $I(hkl)$. Fortunately the effects of such an omission bear much more strongly on the values of the thermal-vibration parameters for the atoms in the final refined model of the structure than on the positional coordinates, so that structural analyses performed without absorption corrections remain valid, although less accurate.

If a four-circle diffractometer is used, all the $I(hkl)$ are collected on a common intensity scale, assuming equal source intensity. This does not apply to different levels in Weissenberg geometry. Furthermore, data from separate

TABLE 3.1

Examples of the range of transmission factors for X-rays

Compound	μ/cm^{-1} for Mo K_α	Crystal dimensions/mm	T
$H_2Ru_6(CO)_{18}$	33·15	$0·20 \times 0·63 \times 0·44$	0·362–0·632
$[(CH_3)_4N^+]_2Fe_6(CO)_{16}C^{2-}$	24·38	$0·20 \times 0·60 \times 0·06$	0·348–0·872
$Ru_3(CO)_7C_{10}H_8$	24·85	$0·24 \times 0·04 \times 0·32$	0·582–0·905

References for these three structures are given in the Bibliography (p. 95).

Weissenberg photographs is affected by exposure times and the sensitivity of different pieces of film. This difficulty is overcome by cross-correlation of intensities collected from crystals mounted along different axes. Thus the levels $hk0$ and $hk1$ can be scaled together with the aid of reflections from both sets appearing in $h0l$. Lastly, the full set of $F(obs)$ must be expressed in electrons, in relation to the total electron content, and thus scattering power, of the contents of the unit cell. Since the elemental composition of the crystal is usually known from chemical analysis, one can use the atomic form factors to predict the expected mean diffracted intensity as a function of Bragg angle. The *Wilson plot* of observed $F(hkl)$ against $(\sin\theta/\lambda)$, based on these considerations, provides an estimate of the scale factor that must be applied to $F(obs)$ and of the mean thermal vibration amplitude of the atoms.

Elucidating the structure

Armed with his set of $F(obs)$, indexed by (hkl), properly scaled, and stored on paper tape, cards, magnetic tape, or a magnetic disc file, the crystallographer can now tackle the solution proper of the structure. The weapons used are those described in Chapter 2; handling them involves the use of a set of computer programs to carry out the various lengthy calculations required. These programs offer great advantages but also some perils. The advantages hinge mainly on the factor of speed: if no difficulties arise in solving the structure, it is often possible to complete it within a few weeks. Should difficulties arise, the computer becomes proportionately more valuable, since false starts are no longer a crippling waste of time and effort. Some determinations are by no means straightforward, even for small structures, and it is a great help to be able to experiment with the different techniques for assigning phases to reflections. Structure-factor calculations are a matter of routine; Fourier maps are produced by numerical sampling of electron density at specified intervals along each cell axis; least-squares refinement depends on the use of a computer, since it is very calculation-intensive; follow-up operations, such as calculations of molecular geometry, are greatly speeded. The risk is that this facility may engender a 'black box' mentality towards computer programs, which are often long, complex, and extensively modified hand-me-downs. Nothing is more disconcerting than to discover, too late, that a least-squares program contains a subtle error buried somewhere in its highly complex interior. The wise man tries to be aware of what his calculations actually achieve at all stages, in order to keep a firm grasp on the credibility of the results.

Most experimental techniques are soon pressed to the limit of their capabilities. The length of calculations no longer poses the same drastic restriction on the size and kind of structure that may be tackled; nevertheless, the computational problems do grow rapidly with size of structure. Within a given limit of resolution, the number of Bragg reflections that may be collected is proportional to the unit-cell volume. The unit cell contains at least one asymmetric unit, which in turn is at least part of the basic chemical unit of the compound being studied. The larger, more complex, and less symmetrical that chemical unit, the larger the unit cell. In addition, the phase problem becomes most severe with large organic structures, such as proteins. The general trend will be illustrated by reference to three structures of increasing complexity, drawn from the experience of the author in the field of organometallic chemistry.

Bisfulvalenedi-iron has the structure **1**. This was suggested by spectroscopic and analytical data but an X-ray diffraction study was undertaken to prove it beyond doubt. The unit-cell volume was 0.7072 nm^3 and 1162 independent, non-zero reflections were collected. The space-group symmetry was such that each molecule of $C_{20}H_{16}Fe_2$ had an exact centre of symmetry; the asymmetric unit was therefore $C_{10}H_8Fe$. This small, simple structure was routinely solved from the Patterson map; some two man-months and one hour of time on a large computer installation were required. By contrast the newly-synthesized copper cluster complex $[HCuP(C_6H_5)_3]_6$ had a unit-cell volume of 19.69 nm^3 and yielded 8905 reflections. Over $32\,000$ might be expected on the basis of

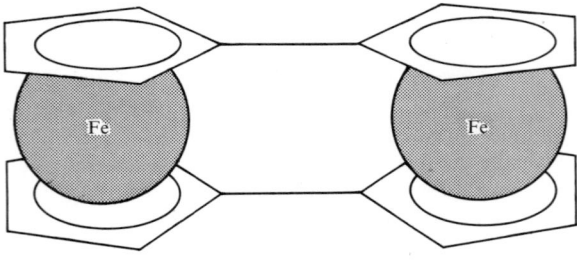

1

relative unit-cell volumes, compared to the previous compound; however, this unit cell was orthorhombic instead of monoclinic, which effectively doubles its symmetry and halves the number of independent families of planes. In addition, the poor quality of the crystals effectively halved the attainable resolution. The limit of ($\sin \theta/\lambda$) for useful data was 0·40, as opposed to 0·59 for bisfulvalene-di-iron. The lack of spectroscopic data on the compound, and particularly the failure to obtain a reliable molecular weight, rendered the full structural analysis indispensable. It also forced the use of direct methods to identify the heavy-atom core of the molecule, shown in Fig. 3.6. This still left the 108 carbon atoms of 18 phenyl groups to be located; a molecule of solvent of crystallization (dimethylformamide) was also found at a late stage in the analysis. No hydrogen atoms could be located and the positions of the hydride ligands remain open to speculation. The process of finding all the carbon atoms, so straightforward in the previous case, turned out to involve a three-month-long ordeal of staring at Fourier maps, repositioning atoms, attempting partial least-squares refinements, and starting all over again. Atomic parameters could only be handled in groups; otherwise the least-squares refinements exceeded the capacity of even a very large computer (IBM 360/65); each refinement cycle took fully twenty minutes on this large and fast machine. A chemically sensible final structure was obtained, although carbon–carbon lengths were very uneven. Most of the problems arose from the sheer size of the structure and the poor quality of the reflection data. With more than 100 atoms in the asymmetric unit, the analysis leaves the realm of the

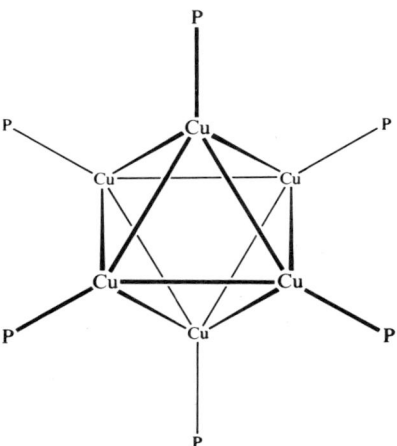

FIG. 3.6. The heavy-atom core of the complex $H_6Cu_6(PPh_3)_6$.

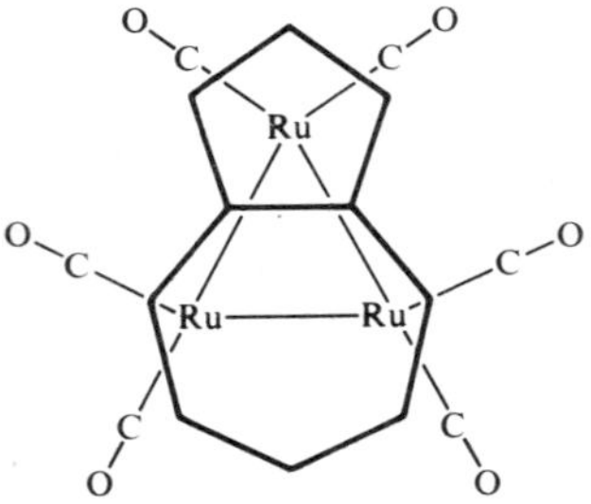

FIG. 3.7. Azulenetriruthenium heptacarbonyl. The seventh carbonyl ligand is obscured.

routine determination and borders on a major battle. When real difficulties arise in elucidation, a long period of repeated attacks may be needed. Azulene-triruthenium heptacarbonyl (Fig. 3.7) was such a case. The Figure merely shows the molecule of the compound. The isosceles triangle of three heavy atoms ought to produce a readily discernible pattern in the Patterson map. Unfortunately, the unit cell dimensions, space group and density required a molecular weight of about double that of the molecular formula $Ru_3(CO)_7C_{10}H_8$. The mass spectrum showed the presence of two distinct species $Ru_4(CO)_9C_{10}H_8$ and $Ru_2(CO)_5C_{10}H_8$. The Patterson map was extremely confusing and direct methods only added to the mess by repeatedly suggesting the presence of a cluster of *four* ruthenium atoms. Eighteen months of intermittent head scratching and experience with a similar but even worse structure finally yielded the clue that there were *two* chemically identical molecules of $Ru_3(CO)_7C_{10}H_8$ in the asymmetric unit, so oriented that their patterns in the Patterson map were completely intertwined. The two species in the mass spectrum appear to result from the disproportionation reaction:

$$2\,Ru_3(CO)_7C_{10}H_8 \quad \rightarrow \quad Ru_4(CO)_9C_{10}H_8 \;+\; Ru_2(CO)_5C_{10}H_8.$$

All very simple in retrospect but a fair proof that crystallography can still be something of an art or knack!

Crystallographic results

Once refinement is complete, the results must be written up, hopefully for publication. This section is intended as a brief guide to those who find themselves having to read crystallographic papers. These usually contain a good deal of information, which should be examined with some care. Most full papers begin with a summary of essential details. A preamble describes the purpose of the determination. The experimental section and that on elucidation and refinement follow; these can give a valuable indication of the care taken in data collection and processing—the better the data, the better the possible result. A list of F(obs) and F(calc) is often provided as a permanent record of the original experimental data and can be examined for particularly bad mismatches, although the most outrageous are generally removed at an early stage, provided some adequate justification can be found. A list of the coordinate positions and identities of all atoms in the asymmetric unit is always provided and from these interatomic distances and bond angles are calculated. To list all of them is usually unnecessary, if not impossible for reasons of space; authors of papers are successful to a varying degree in picking the significant dimensions and presenting them in a rational manner. The discussion of these parameters and the interpretation of their chemical significance follow. If there is doubt, it may be helpful to perform one's own calculation of distances and angles from the list of atomic coordinates. Well-planned drawings are worth a thousand verbal descriptions of the kind 'atom A is bonded to atom B at a distance of x pm; B is bonded to C at y pm' and so on *ad nauseam*. Plotting devices attached to computers make this a relatively easy task and it is therefore surprising how feeble many illustrations actually are. Conversely, some authors manage to produce drawings that are real works of art.

Much of the discussion of structural data consists of comparisons of dimensions, both within the same structure and with other structures. Now the refinement process involves the juggling of the parameters that define the model, with the aim of minimizing the discrepancies between F(obs) and F(calc). This minimization covers all the pairs of reflections at once and is therefore a statistical process which leads to associated uncertainties in the final values of the parameters. Least-squares programs are therefore made to produce not only the values of these parameters but also estimated standard deviations for them, or esd's for short. While the final R-factor is a good guide to the overall fit of the model to the experimental data, close inspection of the esd's is a more severe and preferable test. No comparison of two or more structural parameters is valid without knowledge of the esd's for them.

PROBLEMS

3.1. Why are the rotation photograph and the $h0l$ Weissenberg photograph, both taken from a b-mounting, inadequate for determining the unit-cell parameters for a triclinic lattice?

3.2. A crystal is mounted along its b axis, which corresponds to a unit-cell axis of 1000 pm. A rotation photograph is taken with Mo K_α radiation (71·07 pm), using a cylindrical film of 60 mm diameter. Measuring from the 0th layer line, how far out will the 1st, 2nd, and 3rd appear?

3.3. You have no computer program for calculating absorption corrections but you want to collect the most accurate intensity data possible under the circumstances. You know the linear absorption coefficient, μ, for your crystal, and the transmission factor varies as $\exp(-\mu x)$, where x is the path length through the crystal. Scattering intensity is proportional to crystal volume. Statistically, the stronger a reflection the more accurately it can be measured. What would you do to optimize the results of your experiment?

3.4. Atomic volume increases very much less rapidly than atomic number through the periodic table. Bearing in mind the interference principles outlined in Chapter 2, which part of a set of reflection data is likely to give you the greatest help in locating hydrogen atoms?

3.5. A crystallographer quotes his unit-cell dimensions as $a = 1000 \pm 1$, $b = 2000 \pm 15$, $c = 1500 \pm 15$ pm, for an orthorhombic unit cell. He collects his intensity data very carefully, using a good diffractometer, solves the structure, and refines it to a very low R-factor. Examining the molecular geometry, he finds two carbon–carbon single bonds, 154.2 ± 1.1 and 150.1 ± 1.0 pm long. Is he entitled to postulate any chemical difference between them?

4. The importance of structural analysis

The growth of X-ray crystallography

BEFORE proceeding to specific examples of its usefulness, let us examine the historical growth of X-ray crystallography. The rate of publication of complete structure determinations for carbon-containing compounds is plotted against time in Fig. 4.1. The resulting curve shows two striking features: a very long induction period and a subsequent rapid rise. The first part clearly resulted from the laborious nature of complete structural determinations during the formative years of X-ray crystallography; the second is quite obviously associated with the introduction of high-speed digital computers. The whole curve graphically summarizes the evolution of crystallography from a specialized science into a widely used technique. It even shows signs of turning into the well-known S-curve for the sales of a new product—perhaps market saturation is being approached for the services of crystallographers!

A breakdown by areas of chemistry is also interesting. The data used to compile Fig. 4.1 has been broken down as follows (Kennard and Watson, 1970):

Organic compounds	43·1 per cent
Metal coordination complexes	22·9 per cent
Biological and natural products	15·2 per cent

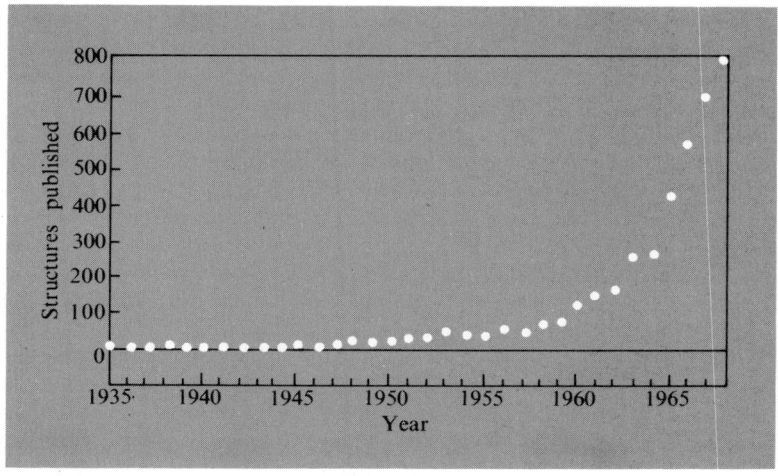

FIG. 4.1. The rate of publication of carbon-containing structures, 1935–1968.

Metal π-complexes   7·0 per cent
Molecular complexes   3·5 per cent
Others   8·3 per cent

The 'growth markets' at present are in coordination chemistry, organo-metallic compounds, and natural products. Increasingly rapid development is likely in organic compounds and biological structures, although the latter are usually very large and involve a different scale of effort and time. The historical or 'mature' markets are in inorganic structures, minerals, and metals and alloys. A survey of ten widely read American, British, Russian, and Scandi-navian chemical journals, which contain the bulk of new single-crystal X-ray determinations, revealed 81 of these in January 1972 alone. At this rate, the annual output must be approaching 1000. This is certainly a big business and we shall now examine the areas on which it has had and is currently having an impact.

Inorganic chemistry and mineralogy

Inorganic chemistry and mineralogy are vast fields and, for the purposes of this section, we shall adopt a restrictive definition of their scope. The discussion will focus on those compounds which, while having stoicheiometric composi-tion, do not involve discrete molecules. We are therefore concerned here with ionic lattices, infinite covalent arrays and states intermediate between these. Many of the compounds are of mineral origin and range from the very simple ionic lattice of sodium chloride to the rich variety of covalent networks found in the silicates. There are thousands of different structures and we cannot attempt to classify them in so little space. The richness is not completely random, however, and inorganic chemists have been able to isolate clear structural trends and relate these to the chemical properties of the con-stituent elements. The classic reference in this field is '*Structural inorganic chemistry*', by A. F. Wells (1962). This very long, exhaustive, and authoritative work of reference has been supplemented by the same author's monograph '*The Third Dimension in Chemistry*' (1965), in which space-filling by geometrical bodies and networks is analysed and the theoretical possibilities are related to the structural realities, particularly in the last two chapters (pp. 87–133). At the very end, Wells pays tribute to X-ray crystallography as the source of much of the structural data. In this section, we shall confine ourselves to examining quite simple examples of structural data and the insights into the mechanisms of chemical combination which they have afforded.

Every chemist learns at an early stage that sodium metal and chlorine gas react violently together in stoicheiometric proportions to produce sodium chloride. This must be the pure chemical compound with the longest human associations, on account of its very wide natural distribution and essential part in nutrition. Its other outstanding characteristic is solubility in water to give conducting solutions. Despite its ready solubility, common salt obviously has

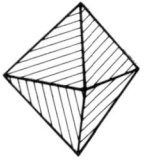

Fig. 4.2. The rare octahedral crystalline form of sodium chloride. The commoner form is cubic.

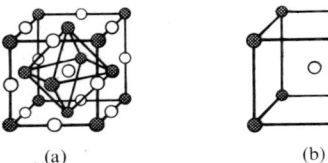

(a)   (b)

Fig. 4.3. The ionic structures of (a) sodium chloride, (b) caesium chloride.

strong structural features as a solid, as witnessed by the distinctive shape of its crystals, shown in Fig. 4.2. The characteristic octahedral shape turns out to be intimately bound up with the internal structure, which is shown in Fig. 4.3(a). There are in fact two interpenetrating face-centred cubic (f.c.c.) lattices, one of Na^+ and the other of Cl^- ions. The result is that each cation is surrounded by an octahedral arrangement of six anions, and vice versa. We therefore say that sodium chloride exhibits *octahedral* 6:6 *coordination*. For comparison, the structure of caesium chloride appears in Fig. 4.3(b): this is a case of *cubic* 8:8 *coordination*. Why the difference? Crystallographic data supply the raw evidence and, on further study, a convincing explanation.

As a first step, a model of the system is required and the generally accepted form, implicitly assumed above, is of spherical cations and anions held together by electrostatic forces. This is quantitatively satisfactory, as shown by the *Born–Haber cycle* of Fig. 4.4. We can measure the heat of vaporization of the

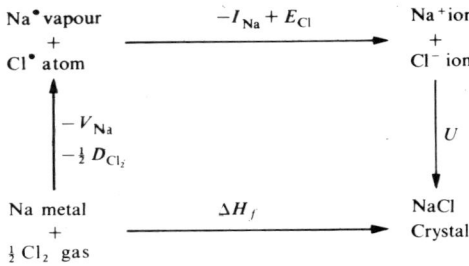

Fig. 4.4. Born–Haber cycle for sodium chloride.

sodium metal, V_{Na}, and the heat of dissociation of the chlorine molecule, D_{Cl_2}, calorimetrically, as also the heat of formation of NaCl from the elements, ΔH_f. The ionization potential of sodium, I_{Na}, and the electron affinity of the chlorine atom, E_{Cl}, come from spectroscopic data. From these and using eqn (4.1), we can determine the lattice energy, U. Theoretical values of U, in

$$\Delta H_f = U - V_{Na} - \tfrac{1}{2}D_{Cl_2} - I_{Na} + E_{Cl_2} \qquad (4.1)$$

good agreement with experiment, can be calculated on the basis of a purely electrostatic model through eqn (4.2). The quantity A is the Madelung constant,

$$U = e^2 z^2 LA(1 - 1/n)/r \qquad (4.2)$$

where e = electronic charge
  z = highest common factor of the charges on the two ions
  L = Avogadro's number
  r = distance between the nuclei of two unlike ions
  n = an empirical integer, ca 10, which depends on the compressibility of the ions

which is a purely geometrical factor governed by the coordination environment of each ion. Our electrostatic model, therefore, is a good one and we can accept the notion of Na^+, Cs^+ and Cl^- as charged spheres.

The critical difference is in dimensions; r_{NaCl} is 279 pm and r_{CsCl} is 346 pm. By itself, this in uninformative but when this dimension was examined for compounds containing combinations of a number of different cations and anions, cross-correlation made it possible to establish a table of *ionic radii*, which remain relatively invariant between different salts. The Goldschmidt scale gives these values: $Na^+ = 98$ pm, $Cs^+ = 165$ pm, $Cl^- = 181$ pm. Now we can think more critically about the coordination patterns. A cation will collect around itself as many anions as it can manage, until these anions come into contact with each other. The larger the anions, compared to the cation, the fewer can surround the cation in its coordination sphere. The results of comparative studies are summarized in the *radius-ratio rule*, expressed in the data of Table 4.1. The rule is remarkably successful in predicting the coordination patterns in purely ionic compounds—so much so that deviations from it can be used as a test of departure from purely ionic behaviour. Its application is discussed in greater detail by Addison (1961).

Structural studies have helped us rationalize the influences controlling the choice of lattice for distinct, monatomic ions. They have been equally important in forming our understanding of infinite covalent lattices. Let us, as an example, examine the silicate minerals. Silicon and oxygen are the two commonest elements in the earth's crust—a fact dependent on the stability and inertness of the structures they form in combination. Silicon itself has the diamond structure, an infinite network of atoms covalently bonded in tetra-

TABLE 4.1

The radius-ratio rule for ionic compounds

r_X/r_A	Usual coordination number about A	Configuration
0·155–0·225	3	Triangle
0·225–0·414	4	Tetrahedron
0·414–0·732	6	Octahedron
0·732–1·000	8	Cube

hedral 4:4 coordination. Crystalline silicon is a hard material, with a high melting point, and structurally similar to diamond. The dioxides of carbon and silicon are utterly different: the former consists of discrete molecules, in CO_2 gas; the latter preserves the tetrahedral coordination environment of solid silicon but with an oxygen atom inserted into each bond. There are various forms of silicon dioxide, silica, SiO_2, found in nature. In β-crysto-ballite, the oxygen atoms are almost linearly and equidistantly interposed between silicon atoms, thereby having roughly linear 2:2 coordination forced upon them, in contrast to the bent arrangement they usually favour. In β-quartz, this strain is relieved to some extent and the bending of Si—O—Si bonds produces the helical structure and characteristic optical activity of that mineral.

Rather few structures containing monomeric or dimeric silicate anions are known; the only two containing the cyclic trimer $[Si_3O_9]^{6-}$ are the clay minerals benitoite, $BaTiSi_3O_9$, and wollastonite, $CaSiO_3$. The $[Si_6O_{18}]^{12-}$ anion occurs in beryl, $Be_3Al_2Si_6O_{18}$, whose structure is shown in Fig. 4.5. Beryl gemstones are coloured golden-yellow by the presence of transition-metal cations or impurities; emerald and aquamarine arise in a similar way. Large synthetic emeralds can be made by hydrothermal synthesis. The amphiboles, whose general structure is shown in Fig. 4.6, incorporate hydroxyl groups; by this device, the net charge per silicon atom is directly lowered by comparison with the pyroxenes which have a similar structure without these groups. Fewer countercations are therefore needed in the amphiboles and the chains of condensed SiO_4 tetrahedra are only weakly held together by hydrogen-bonding. Asbestos, although formally containing the $[Si_2O_5]^{2-}$ anion, is more complex and really an amphibole; the mineral readily separates into fibres which are much used as reinforcements in composite materials, and especially when fire resistance is required. Whereas amphiboles have two 'free' or unshared apices for each tetrahedral SiO_4 unit, micas and talcs have

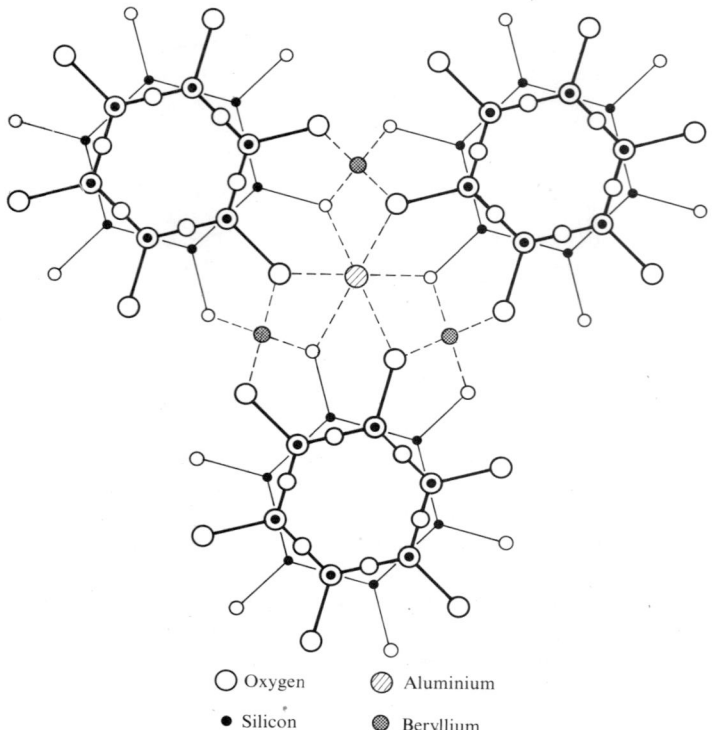

Oxygen

Silicon

Aluminium

Beryllium

FIG. 4.5. The structure of beryl. The $[Si_6O_{18}]^{12-}$ rings are stacked above each other, alternatively staggered, and the Be^{2+} and Al^{3+} cations fit into the channels between the stacks.

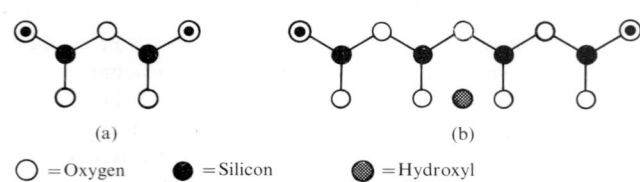

(a)

(b)

$\bigcirc$ = Oxygen $\bullet$ = Silicon $\bigcirc$ = Hydroxyl

FIG. 4.6. Pyroxene (a) and amphibole (b) chains in cross-section. The chains pack alternately up and down, with cations between them.

but one. Talc is in fact the condensation product of the acid anion $[Si_2O_5]_n^{2n-}$ and magnesium hydroxide, with the formula $Mg_3(OH)_2Si_4O_{10}$. Its structure accounts for its slippery feel, which is a result of the weakness of the forces holding the silicate sheets together—a situation akin to that found in graphite and molybdenum disulphide. Apart from silica itself, the three-dimensional nets are based on the partial replacement of silicon atoms by aluminium. This tends to create a more open structure and because an SiO_4 tetrahedron is replaced by AlO_4^-, countercations must be introduced. In the zeolites, of which one is shown in Fig. 4.7, large, regular voids and interconnecting passages are created; these compounds are therefore used as molecular sieves and ion-exchange agents.

This series of compounds reveals, through the study of their structures, many of the features that we have come to associate with covalency. First and foremost is the tendency of the silicon atom to maintain a tetrahedral environment about itself, through appropriate hybridization of its valence orbitals. High, localized charges are avoided by the formation of condensed (polymeric) structures.

The acid–base behaviour is also reflected in these condensed structures. Silica and the silicates span the gap between the purely ionic structures formed by elements of widely differing electronegativity, and the strictly covalent networks, consisting of atoms of very similar electronegativity. Closer examination of the structural details allows the assignment of *covalent radii* to atoms, by division of bond lengths into individual contributions. The relationship between covalent and ionic bonding, electronegativity, and bond lengths and strengths is fully explored by Coulson (1961 and 1973). On the practical level, we obtain a much clearer understanding of the chemical and physical properties of the various silicates.

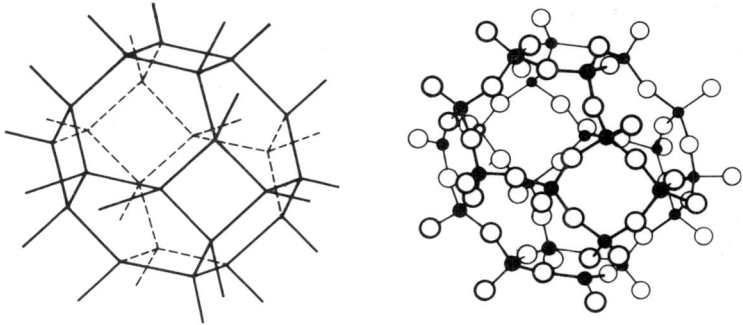

FIG. 4.7. The basic skeleton of the ultramarines and the synthetic Zeolite A.

Metal complexes and covalent molecules

At one extreme of the ionic–covalent scale lies sodium chloride; it readily forms solutions which show not even a vestigial trace of the solid structure of the salt, except at extremely high concentrations. At the other end of the scale, the infinite covalent lattices will not dissolve at all. Much research has, however, been concentrated on the *coordination compounds* or *complexes*, in which *ligands* are closely bound to a metal atom or ion, and remain so, whether in the solid or in solution. These are mainly found for the metals of the transition series. A familiar example is the ammonia–copper(II) cation, which is really a mixture of cations in aqueous solution, ranging from $[Cu^{II}(H_2O)_6]^{2+}$ to $[Cu^{II}(H_2O)_2(NH_3)_4]^{2+}$. The fifth and sixth water ligands can be replaced by ammonia ligands in liquid ammonia solution to yield $[Cu^{II}(NH_3)_6]^{2+}$. A great deal of work has been put into determining the equilibrium constants of complex formation in solution, the complexing abilities of various ligands, and the reactivities of the complexes so formed. Structural determinations have played an important role in the characterization of complexes and in the theoretical treatment of their stabilities.

Because many complexes are at least kinetically stable in solution, Werner and Jørgensen—the founders of coordination chemistry—were able to investigate them by chemical means. Werner in particular studied isomerism in coordination complexes and was thus able to postulate structural formulae, dealing mainly with octahedral Co(III) and square-planar Pt(II) species. An account of his work and of the developments to which it has led may be found in the Werner Centennial volume (1967). The paper contributed by R. G. Wyckoff to this symposium (pp. 114–119) on Werner coordination and crystal-structure analysis is especially illuminating and brings out the extent to which chemistry is a geometrical science—a view much developed by Wells in his books. Werner's postulated octahedral and square-planar structures

$$\left[\begin{array}{c} NH_3 \\ | \quad NH_3 \\ H_3N \!-\!\! Co \!-\! NH_3 \\ | \\ H_3N \quad NH_3 \end{array} \right]^{3+} \quad \left[\begin{array}{c} NH_3 \\ H_3N \!-\! Pt \!-\! NH_3 \\ H_3N \end{array} \right]^{2+}$$

have been confirmed and, as X-ray diffraction has become an increasingly routine instrument of characterization, many instances of three-, five-, seven-, and other less common coordinations have been demonstrated.

Transition-metal complexes have provided a fertile field in which to cultivate theories of bonding. These range from the simple electrostatic *crystal-field theory* to the highly sophisticated *molecular-orbital theory*, based on covalently bonded ligands. Predictions based on either of these theories require as input accurate structural parameters; the accumulation of structural data has provided a series of phenomena to be catalogued and explained. The electronic configurations of the transition metals, with their partially filled d-shells, lead to variations of geometries and bond lengths in their complexes. The properties of ligands and the coordination geometries that they induce are beginning to be understood in some detail.

This knowledge is now being exploited in the construction of complexes intended to model the catalytic activity of biological metal complexes, which act in a very potent and specific manner. Transition-metal complexes have always been used, albeit in an empirical fashion, for industrial catalysis. Fundamental studies of the structures of these catalysts offer the prospect of ultimately tailoring the complexes to achieve the desired effects, and this has been a powerful stimulus to research in this area. As a brief example, we may quote work on metal–porphyrin complexes, which are actively being investigated as model systems for nitrogen fixation. A recent publication (Fleischer, 1972) describes the reduction, at 1 atm pressure and room temperature, of acetylene and atmospheric nitrogen by sodium borohydride and the complex shown in Fig. 4.8. This system achieves the fixation of atmospheric nitrogen under essentially the same conditions as the enzyme nitrogenase, which is responsible for the same reaction in living organisms. These organisms perform, under very mild conditions, one of the most important chemical processes in nature; we can only parallel it under the brutally forcing conditions of the Haber process. To mimic the work of an enzyme, however crudely, has been a major goal of recent chemical research and constitutes a substantial achievement. Porphyrins and metalloporphyrins have been closely characterized structurally because they form the cores of several important

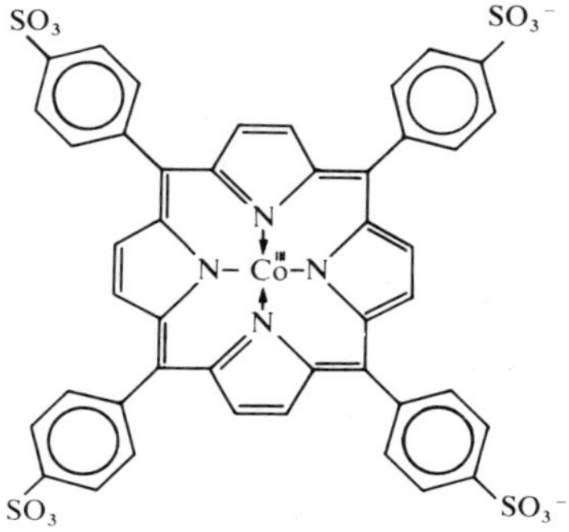

Fig. 4.8. The complex *meso*-tetra(*p*-sulphonatophenyl)porphinatocobalt(III).

biological catalysts, of which three notable examples are chlorophyll, haemo-globin, and vitamin B_{12}. X-ray diffraction studies have been extensively used in examining the nature of the bonding in these ligands and complexes. The important parameters measured include bond lengths in the ligand, which reflect the nature of its aromaticity, the nitrogen–metal distances in the metalloporphyrin complex, which are guides to the strength of complexation of the metal, and the coordination geometry of the metal ion, which may be correlated with the reactivity and catalytic activity of the complex. The contribution of diffraction studies to our understanding of this area of chemistry has been described in a short review article by Fleischer (1970).

Compounds containing the smaller covalent molecules of the non-transition elements are often gaseous or vaporizable, or may be soluble in organic solvents. They can therefore readily be studied by spectroscopic techniques, which are greatly bolstered by the application of symmetry rules. Infrared, nuclear magnetic resonance, electron spin resonance, and other techniques are routinely applied to small molecules. Diffraction techniques are useful when precise geometrical data are required for the elucidation of the more subtle bonding effects. Single-crystal X-ray diffraction is a logical technique for use with solids; vapours or liquids can be frozen and studied at low temperatures—usually by cooling with a stream of cold nitrogen—but it is often preferable to examine the molecules in the normal physical state of the compound. Not much can be done with liquids but vapours may be studied by means of electron diffraction, as explained in Chapter 7. An interesting series of species, of which many have been studied by X-ray diffraction, are the *carboranes*. An example is shown in Fig. 4.9 and the reader is referred to a review article by Hawthorne (1968) for further descriptions and a discussion of these derivatives of the boron hydrides, which can also be converted into ligands for transition-metal atoms and ions.

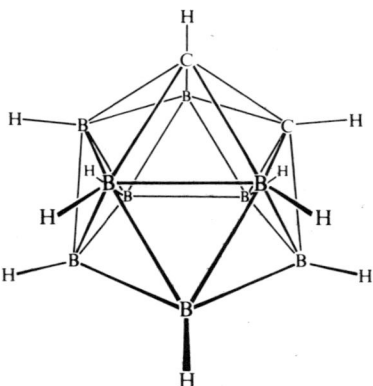

FIG. 4.9. $1,2\text{-}B_{10}C_2H_{12}$, an icosahedral carborane.

Organometallic compounds

The study of bonding between metals and carbon atoms is one of the fastest-growing areas of chemistry. It is also a recently developed one, the field having first been seriously investigated after the discovery of the novel compound ferrocene (2) in 1951–2. The synthesis of a multitude of organo-

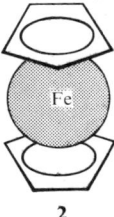

2

metallic compounds has offered new perspectives in chemical bonding and structural determinations have been crucial in shedding light on the possible modes. For non-transition metals the primary mode is that of σ-bonding, which involves a conventional electron-pair covalent single bond between the metal and a carbon atom. The compounds so formed are often volatile; some have great commercial importance, notably tetraethyl- and tetramethyl-lead. A few compounds contain a metal cation and an organic anion.

There still exists something of a division between the organometallic chemistries of the main-group and transition-series elements; the latter area has proved much the richest, at least in terms of structural concepts. As always, spectroscopy and analysis have facilitated the rapid screening of new compounds that must take place in a field of great synthetic activity—hundreds of new compounds are created every year. Characterization by these techniques is, however, invariably made within the reference framework of the structural principles of organometallic chemistry. Our now considerable stock of structural information can be rationalized in various ways; we shall briefly survey it in an order dictated by the number of electrons participating in the metal–ligand link.

The simplest case is the straight σ-bond, which involves a single electron from the ligand. An interesting example of this is tetrabenzylzirconium (3). A

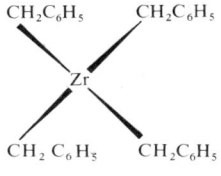

$$CH_2C_6H_5 \qquad CH_2C_6H_5$$
$$Zr$$
$$CH_2C_6H_5 \qquad CH_2C_6H_5$$

3

good deal of thought has been expended on the factors governing the stability of metal–carbon σ-bonds : the subject is by no means simple, since the stability appears to be considerably influenced by the nature of the other ligands attached to the metal. Tetrabenzylzirconium, for example, is quite stable but has the interesting property of catalysing the polymerization of ethylene, apparently through the insertion of successive molecules of monomer into the single bond.

Next comes the involvement of two ligand electrons, typified by Zeise's salt (4). This compound, dating from the nineteenth century, shows a quite different

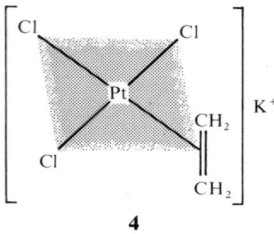

4

mode of bonding, the metal–ligand π-bond. It is now generally agreed that this π-bond is a composite bond, in which electrons from the π-orbital of the ligand flow into low-lying metal orbitals, with a reciprocal *back-donation* from more energetic metal d-orbitals into the vacant π*-antibonding orbitals of the ligand. There is still constant argument about the extent of the back-bonding contribution and the manner in which its strengthening of the overall bonding is reflected in a contraction of the metal–ligand distance. The phenomenon is also seen as a form of double-bonding, superimposed on more conventional σ-bonds ; thus it has been argued that the $M-C_3F_7$ bond is shorter than the $M-C_3H_7$ bond in otherwise similar complexes because of the greater electronegativity of the fluorinated ligand, which draws π-density into its antibonding orbitals.

In the very common π-allyl compounds, such as (**5**), three ligand electrons

$$
\begin{array}{c}
\text{CH} \\
\text{CH}_2 \quad \text{CH}_2 \\
\text{Mn} \\
\text{C} \quad \text{C} \quad \text{C} \quad \text{C} \\
\text{O} \quad \text{O} \quad \text{O} \quad \text{O}
\end{array}
$$

5

are implicated, whereas four are involved in the butadienyl structures, such as (6). Ferrocene can be considered as having five-electron donor ligands, $C_5H_5^{\bullet}$,

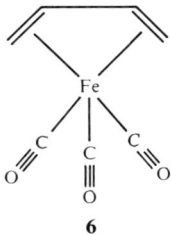

6

or six-electron donors, $C_5H_5^-$. Uranocene (7) contains two planar π-cyclo-octatetraenyl ligands, each donating eight electrons.

7

One should remember, however, that these exercises in counting electrons are to some extent formalized descriptions. Formalisms are useful, though, in rationalizing the evidence before us; one of the simplest and most widely used is the Effective Atomic Number of EAN rule, according to which the metal atom strives to achieve the electronic configuration of the next noble gas. As an example of the importance of detailed information on bond lengths and angles, consider structure (8): this is merely (6) writ different—the electron

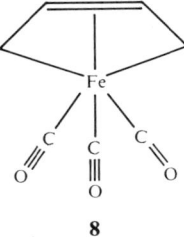

8

count remains the same, as does the oxidation state of the iron atom. On the other hand, the bond orders in the butadiene ligand have been reversed, so that the central carbon–carbon bond should be shorter than the outer two in (8) but longer in (6). In actual fact, X-ray data shows rather little difference between the two lengths, so that the true position may be presumed inter-mediate between (6) and (8).

The structures of organometallic compounds have stimulated much theoretical work on modes of bonding, notably the forward- and back-dona-tion concept. Considerations of orbital symmetry are very useful in this context, since they provide a definitive qualitative criterion of what is or is not possible. One interesting viewpoint is that bonding to transition metals can stabilize the excited states of organic ligands. Thus fulvalene (9) and cyclo-butadiene (10) have, at best, fleeting existences as ground-state organic mole-cules but have been isolated in their excited states in the complexes (1) (p. 30) and (11). Mason and Robertson (1966) have written an account of the inter-

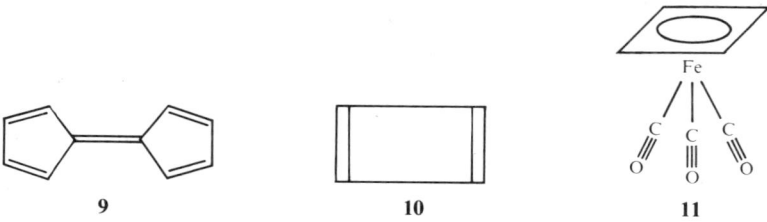

action of diffraction methods and quantum chemistry. There is also an apocryphal story of a famous theoretician who was consulted on the matter of ferrocene, performed a rapid calculation on one side of a cigarette package, and declared it unstable. Later presented with all the evidence of a highly stable compound, he did another calculation on the other side of the same package and agreed that, yes, it was stable!

Metals and semiconductors

Diffraction studies have revealed that the overwhelming majority of metals have one of three structures illustrated in Fig. 4.10, face-centred cubic (f.c.c.), body-centred cubic (b.c.c.) or hexagonal close-packed (h.c.p.). The only

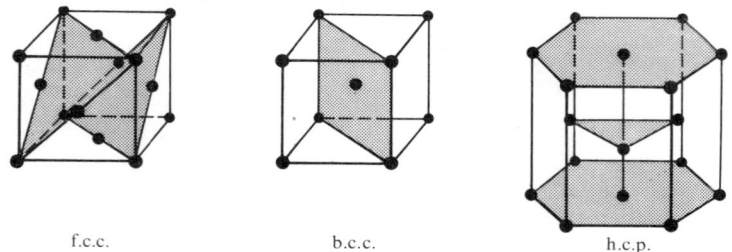

FIG. 4.10. The three common metallic lattices. The packing planes are indicated by shading.

Si (1)

Mn (7) . . . Ga (8)   Ge (1)   As (3)   Se (4,5)
              In (9)   Sn (1,2)   Sb (3)   Te (4)
                                     Bi (3)   Po (6,7)

FIG. 4.11. The less-common metallic structures: 1 = diamond cubic, 2 = body-centred tetragonal, 3 = rhombohedral, 3 = trigonal, 5 = monoclinic, 6 = rhombohedral, 7 = complex cubic, 8 = orthorhombic, 9 = face-centred tetragonal.

certain exceptions are given in the skeleton Periodic Table of Fig. 4.11. The point of this last figure is that it demonstrates the trend away from simple close-packing and towards covalent bonding in the metalloids, such as silicon, germanium, and tin, all of which have the diamond structure.

The metals clearly form a unique structural class exhibiting as they do the third variety of strong chemical bond. The metallic bond is remarkable for its occurrence in such a large variety of elements and electronic configurations. It is, however, the structural similarities rather than the divergences of the metals that are so striking; these similarities are reflected in the physical properties of these elements and to a lesser extent, in their chemical behaviour. All the metals, with the exception of mercury, are dense, lustrous solids at room temperature; all are good conductors. Contrast these common features with, say, the differences between sulphur and chlorine, which are next–door neighbours in the Periodic Table. We know, of course, that the increase in atomic number by one unit between sulphur and chlorine causes an electron to be placed in an exposed and sensitive valence orbital, whereas the orbitals that fill in the metals are much better shielded from chemical interaction. This cannot, however, immediately explain the characteristic metallic properties. All the theories of the behaviour of electrons in metals—for electrons are the glue that holds the atoms together so strongly—require detailed structural data as input for quantitative calculations.

Alloys show a yet wider range of structures and are that much harder to treat. In the simplest binary alloy, atoms of a second metal may replace those of the first at the usual lattice sites, leading to the formation of a *substitutional solid solution*. A good example is the Cu–Ni system, which may contain any proportion of the two metals. In contrast, the composition range of the Cu–Ag solid solution is strictly limited—move outside it and a second phase is precipitated from the hitherto homogeneous solution. In this context, though, homogeneity is a relative term, for whereas the phase as a whole appears uniform, the distribution of solute atoms is locally random and disordered. This is not always so, and the pattern of substitution may in some cases be ordered. Should the periodicity of this ordering exceed the dimensions of the unit cell of the host lattice, then a *superlattice* exists. This effectively consists of larger unit cells, which will therefore give rise to extra Bragg reflections.

Another form of superlattice is encountered in magnetic solids and is discussed in Chapter 6.

Most metals will form intermetallic compounds but definite stoicheiometries are found most commonly in combination with elements of sub-groups IVB, VB, and VIB. The more electronegative—i.e. the less metallic in character—the second element, the more stable and stoicheiometric the compound. Some characteristic examples are: selenides and tellurides of several divalent metals, such as MgSe or PbTe, with NaCl (rocksalt) structure; Mg_2Sn or Sn_2Pt, with the CaF_2 (fluorite) structure; GaAs or InSb, with the ZnS (zinc blende) structure; MgTe with ZnS (wurtzite) structure. Note that the last three classes include metal sulphides, which are really salts; it is therefore arguable whether we are talking of ionic or metallic binding. As the difference in electronegativities decreases, we move to the series of compounds based on the NiAs structure, such as AuSn, CuSn, or PdSb; these show variable stoicheiometry, although within restricted ranges. Finally, short of full mutual solubility, there are many so-called electron phases, or intermetallic combinations, apparently controlled by the concentration of available valence electrons.

Two principal factors control mutual solubility and compound formation: atomic volume and electronic configuration. They are interdependent and any proper theory of the stability of alloy phases must take them both into account.

The role of atomic volumes is perhaps most clearly manifest in the *interstitial compounds*, such as Fe_3C, and the *interstitial solid solutions*. In these, small atoms—notably hydrogen, boron, carbon, nitrogen, and oxygen—occupy the interstices between the close-packed metal atoms. There is invariably some expansion of the unit cell of the host lattice but the possibilities are essentially controlled by a radius-ratio rule. A dramatic example of an interstitial solid solution is the palladium–hydrogen system: palladium metal 'dissolves' hydrogen so effectively—the H_2 molecule splitting in the process—that it can act as a selective filter for that gas. The splitting does, of course, imply that the action is more than purely physical.

Single-crystal diffraction methods may be applied to any single, homogeneous phase and are very valuable for complex alloy structures of this kind. They can often be replaced by powder-diffraction methods (see Chapter 5) for simple, highly symmetrical structures. For either method, cooperative scattering throughout the phase is required and neither can therefore be used when two or more phases coexist, however finely dispersed they may be, if a single unambiguous diffraction pattern is desired. The powder method can, however, be used for the identification of mixed phases whose individual patterns are already known. Even chemically homogeneous phases are likely to contain grains in different orientations, which cause similar problems.

Our best insights into the thermal- and electrical-conductivity properties of metals have come from the *band theory* of electronic behaviour. This will be discussed in a forthcoming book in the series on the electronic structure of solids.

Organic compounds

Because of the early grasp of stereochemical principles achieved by organic chemists, their field of work has, at least until recently, been relatively untouched by diffraction methods. Degradative and synthetic studies have also been powerfully assisted by spectroscopic techniques, particularly n.m.r. (See, for example, McLauchlan, *Magnetic resonance*, OCS 1.) Nearly every organic compound is well-provided with hydrogen atoms which usually afford sufficient n.m.r. resonances with characteristic chemical shifts and fine structure. There is thus little incentive to resort to diffraction methods, unless special information on bond lengths or hybridization is sought, or the problem is especially complex. At the same time, the crystallographic phase-problem is least tractable in organic structures.

More recently, as interest in the biological activity of natural products and synthetic drugs has intensified, the importance of detailed molecular stereochemistry has become increasingly clear. Here, diffraction methods have much to offer and are being used more frequently. Two technical developments have contributed to this trend. The first is the perfection of statistical phasing methods, particularly in their application to the non-centrosymmetric arrangements which organic compounds seem to favour in forming crystals. The second is the determination of absolute optical configuration, through the use of *Bijvoet pairs*.

In our treatment of X-ray diffraction photographs, we stated that the diffraction process adds a centre of symmetry to any other symmetry elements that may be observed. Otherwise stated, this is Friedel's Law, expressed in eqn (4.3). A particular application of this law to the non-centrosymmetric

$$I(hkl) = I(h\bar{k}\bar{l}) \tag{4.3}$$

monoclinic space groups leads us to eqn (4.4), which effectively means that we

$$|F(hkl)| = |F(h\bar{k}\bar{l})| \tag{4.4}$$

cannot find any difference between the intensities of the pairs of reflections (hkl) and $(h\bar{k}\bar{l})$. If, however, the radiation used has a wavelength close to an absorption edge for one of the elements in the compound, then eqn (4.4) breaks down. The intensity difference between these Bijvoet pairs can then be used to determine the absolute sense of the b-axis with respect to the ac plane. Consequently, the absolute sense of the y coordinates of atoms with respect to their x and z coordinates is known, and the configuration of optically active centres can be absolutely related to the crystal axes. The method was first applied by Bijvoet to rubidium potassium tartrate; conveniently enough, it confirmed that the arbitrarily chosen 'handedness', (chirality), to which all other optically-active structures were related by the conventional methods of resolution, was correct. (For a discussion of chirality see Whittaker, *Stereochemistry and mechanism*, OCS 5.)

Biological structures

Natural products and synthetic drugs show important and specific stereo-chemical features. This is, by inference, also true of the living systems with which they interact. Chemical processes in living organisms are indissolubly linked with the structures within which they take place. Complex series of chemical reactions are built up through long chains of interdependent, simple steps, each of which is controlled with amazing precision and efficiency. We have already mentioned the example of the enzyme nitrogenase, which specifically reduces molecular nitrogen to ammonia with an ease that eludes the best chemists. All this points to a very high degree of stereochemical control of reactions in living systems, so that structural analyses form an important part in the study of biological systems at the molecular level. X-ray crystallography is making a substantial contribution in this domain, even though the skill and effort required in solving some of the very large structures are prodigious. Protein crystals, for example, may have unit-cell axes of tens of thousands of pm. The effort expended is rewarded by the flow of detailed information, particularly on the amino-acid sequences in proteins and their precise stereochemistry.

Rather than tackle a huge subject in inadequate space, we shall refer the reader to two illustrative articles, designed to introduce the achievements of biochemical crystallography. The first is by Dickerson (1972) and deals with the evolution and importance of cytochrome *c*, a vital respiratory factor in the higher life forms. A principal theme of the article is the effect of mutations that alter even tiny details of the amino-acid peptide chain; we can only grasp these effects from an exact knowledge of the structure, which single-crystal work alone can supply. Again, the incidence of sickle-cell anaemia as an inherited disease has been traced to a change in a single amino-acid residue in the whole vast complexity of the human body. The determination of large biological structures is being actively pursued in laboratories throughout the world; our second reference (Kartha, 1968), describes the general approach to proteins through X-ray diffraction.

Single-crystal work in perspective

A full-scale single-crystal X-ray diffraction study is perhaps the ultimate weapon in characterizing a chemical species. Like other ultimate weapons, it is expensive and must therefore compete with other, quicker methods. In some circumstances, single-crystal work is manifestly useless—as when the compound stubbornly refuses to yield acceptable crystals, or these decompose too rapidly, or when data are satisfactorily collected but, for some internal reason, the structure simply cannot be solved. Once finished, the crystal structure analysis has a quality of completeness that is hard to match; the chances of misinterpreting the data and producing a totally wrong result are really quite small, given normal standards of competence; the information produced is definite and one can proceed to argue from well-founded facts.

This can, of course, result in a degree of overkill—it is a sheer waste of effort to solve a crystal structure when, say, a good n.m.r. study will answer the question raised. It is also worth remembering that a crystal is, by definition, a solid, and that conclusions valid for the solid state may collapse in solution.

Fluxional organometallic molecules are a case in point. Solid π-$(C_5H_5)Fe(CO)_2-\sigma-C_5H_5$ has the structure shown in Fig. 4.12, as substantiated by the single-crystal work. In solution, however, it shows an anomalous n.m.r. spectrum. At room temperature, a single ^1H resonance is observed for the σ-(C_5H_5) ligand, which might be expected to produce at least two chemically distinct resonances. As the temperature is lowered, the single rather broad peak is resolved into the expected pattern. Detailed analysis of the spectrum as a function of temperature, through computer simulation, shows that this is caused by a 1–2-shift mechanism. In essence, the iron atom appears to hop around the ring, now σ-bonded to one carbon atom, now shifting to its nearest neighbour. As the temperature falls, the shift slows, until the fluxional behaviour is 'frozen out' and the iron atom stays tied to one particular carbon atom. Several examples of fluxional molecules have been observed; they are discussed by Cotton†. Some objections have also been raised to the wholesale application of solid-state data to the behaviour of complex biological systems.

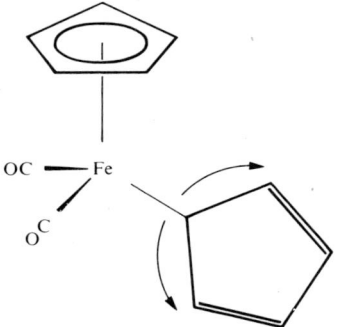

FIG. 4.12. The structure of π-$(C_5H_5)Fe(CO)_2-\sigma-C_5H_5$. The arrows show 1–2 shifts from n.m.r. data in solution.

† F. A. Cotton (1968), *Acc. chem. Res.*, 257. The spectrum of the iron compound above is shown and discussed on p. 260.

Single-crystal work is becoming more and more valuable to the practising chemist, not least because of its great accessibility. There are plenty of crystallographers eager for problems to tackle and it is now far easier for the interested chemist to learn to do his own crystallography. In the wider context of diffraction methods as a whole, single-crystal work does not have the field to itself, even though it may be pre-eminent in the 'purer' fields of chemistry discussed so far. We shall therefore now turn our attention to the wider applications of diffraction methods.

PROBLEMS

4.1. In this chapter, ions are described as spheres. Could you use X-ray diffraction data to test this assumption?

4.2. Ammonium chloride exhibits a λ-transition, or second-order transition, in which its heat capacity appears to rise to infinity over a small temperature range. It has been suggested that this is caused by the onset of free rotation of the ammonium ions. How would you test this interpretation by X-ray diffraction?

4.3. You are given a solution of a pure optical isomer of an optically active compound. Will it crystallize in a centrosymmetric space group, a non-centrosymmetric space group, or in either kind?

4.4. The stable organochromium compound bisbenzenechromium has been prepared. Its structure is a sandwich with the metal atom between two planar hydrocarbon rings, as in ferrocene. Theoretical chemists cannot decide whether the carbon–carbon bond-lengths in the rings ought to be equal or alternating. You undertake a careful diffraction study in order to resolve this point, and your carbon–carbon bond-lengths do not come out exactly the same. What criteria would you apply in reaching a conclusion from your data?

4.5. In a simple metal lattice (i.e. not centred), every alternate atom is replaced by one of a different metal. What additional reflections would you expect to observe?

5. Specialized applications of X-ray diffraction

The powder method

WITH single crystals, the exact orientation of the crystal axes with respect to cameras and diffractometers may be discovered; it is essential to know this for successful data collection, indexing, and analysis. Single-crystal methods, based on the work of Laue and the Braggs, do not have a monopoly of diffraction experiments: a major alternative, which involves no knowledge of crystal orientation, was produced by Debye and Scherrer (and independently by Hull) as early as 1916. The principle of the now widely used Debye–Scherrer camera is shown in Fig. 5.1. The monochromatic X-ray beam enters through the collimator A, bathes the sample B, and leaves through the collimator C, producing a small spot on the fluorescent screen D for alignment purposes. The body of the camera, E, consists of a cylinder, short in proportion to its diameter, centred on the specimen; a strip of X-ray film lines the interior of the camera. The specimen is mounted on a rod, for easy positioning in the X-ray beam, and consists of a powder, either packed into a thin-walled capillary tube or glued to the surface of a fibre. This powder consists of amorphous material, or of a mass of tiny crystallites produced by grinding a sample of crystalline material; the crystallites are assumed to adopt a perfectly random orientation.

Now each crystallite or local element in the powder is able to diffract X-radiation, in accordance with Bragg's Law and under the constraints of the Ewald construction. The powder in effect contains a mass of reciprocal lattices, all jumbled together in different orientations. Statistically, some reciprocal lattice points must lie on the surface of the sphere of reflection. The only restriction on which part of the surface they touch is exercized by the Bragg angle, 2θ. The result is shown in Fig. 5.2, in which the incident beam follows the path QR and the reflected beam RP. P can in fact lie anywhere on the circle described by the constant radius OP, so that the locus of RP is a cone

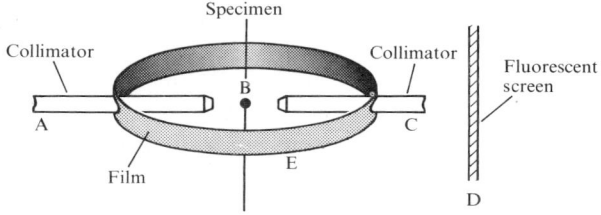

FIG. 5.1. Schematic representation of a Debye–Scherrer camera.

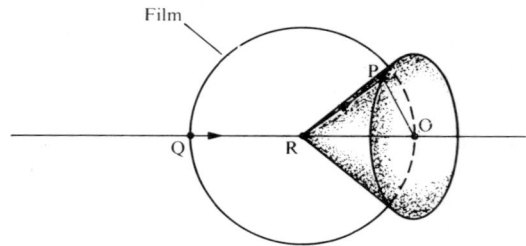

FIG. 5.2. The Ewald construction applied to powder diffraction.

of reflection, whose central axis coincides with the incident beam direction. Each reciprocal lattice point is associated with a length OP, so that a concentric set of cones of reflected radiation is created, as illustrated in Fig. 5.3. The cones intersect the film strip, producing darkening of the emulsion and a pattern shown, for the unrolled strip, in Fig. 5.4. Distance along the film strip from the hole for collimator C is proportional to 2θ. Now we saw that in rotation photographs of single crystals there was an indexing ambiguity between reflections with the same θ, because the photograph could not record the azimuthal orientations of the families of planes about the spindle axis. The ambiguity is compounded in powder photographs: for a cubic structure, the lines for the (100), (010), and (001) reflections exactly coincide. The multiple coincidences that occur inevitably lead to a loss of information, compared to single-crystal data. Nevertheless, good use can be made of what remains,

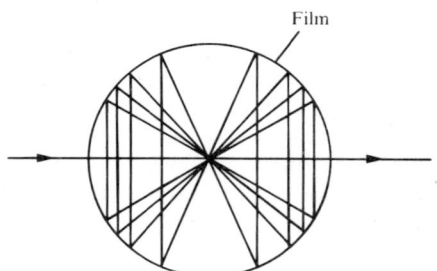

FIG. 5.3. Diffraction cones from a powder specimen.

FIG. 5.4. A Debye–Scherrer photograph.

particularly for lattices of high symmetry; the powder pattern of a cubic lattice is quite easily indexed but monoclinic and triclinic lattices are very hard to deal with. The normal procedure is to guess at some reasonable cell dimensions and compare the observed and calculated powder patterns. If the lattice symmetry is unknown, the best hope is to force-fit a cubic unit cell onto the data and then proceed by successive approximation, lowering the symmetry as necessary. If no headway can be made, then single-crystal methods have to be used.

Line intensities may be read from the film or recorded on a powder diffractometer. This instrument incorporates a counter rotating about the specimen in a fixed plane, in order to vary θ; otherwise, it resembles the powder camera in construction. One advantage of the powder method is that it allows intensification of the diffracted beam through the *divergent-beam* technique of Fig. 5.5: the sample is spread out in a thin sheet and the detector is fitted with a narrow collimating-slit aperture. Once correctly-indexed line intensities are available, structure solutions may be performed in the usual fashion. Overlaps and indexing problems in the lower symmetries greatly hamper the procedure and the powder method has mainly been used in determining the structures of comparatively simple systems, such as salts, minerals, metals, and some alloy phases. Anything more complex requires great ingenuity and a good measure of luck. The result is that, except when suitable crystals cannot be grown, single-crystal methods have largely taken over the field of structure analysis.

The simplified nature of the powder pattern does, however, render it attractive for a different purpose. As it generally contains enough lines to be characteristic of a given set of lattice dimensions but not so many that the pattern is hard to recognize, it serves as an excellent fingerprint for the specimen. Powder-diffraction files have therefore been set up for thousands of different compounds and elements, so that the pattern of any unknown

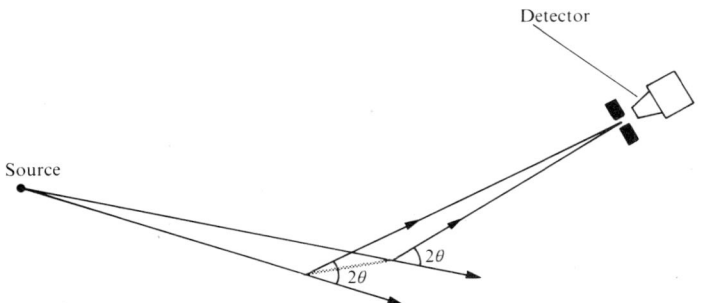

Fig. 5.5. Divergent-beam focussing.

substance can quickly be compared with those in the files. More importantly, the simplicity of the pattern means that it can often be recognized even when it is superimposed upon another; mixed samples can therefore be studied and their components identified. This capability has proved particularly useful in work on the phase-diagrams of alloys: the alloy is made to the required composition and annealed at the appropriate temperature; it is then quickly quenched and ground, or otherwise powdered; the constituent phases at that point of the composition–temperature diagram are identified from the powder pattern. Other techniques, described in subsequent sections, furnish information on the sizes and shapes of the local regions of different phases.

Another area in which the absence of need for sample purity is exploited is that of heterogeneous catalysis. Many heterogeneous catalysts are deposited on solid supports and powder diffraction offers a good way of characterizing the resulting material.

A general factor in favour of powder diffraction is the ease of sample handling, especially in that it eases studies of materials at extreme temperatures and high pressures. In fact, high-pressure diffraction work can really only be attempted by powder methods. Azaroff and Buerger (1958) give a full account of powder methods.

X-ray diffraction and high polymers

The powder method is, as we have just seen, closely related to single-crystal diffraction work; in principle, the crystallites are individually good, single, crystals but the assembly of many of them in random orientations leads to a loss of information. In either case, full three-dimensional order exists at the local level. Our earliest considerations of diffraction gratings suggested that this degree of order is not a precondition for diffraction effects; indeed, order in one dimension only should be enough. If limited information on periodicity alone is sought, then X-ray diffraction has a host of applications. Any structure with markedly anisotropic properties can qualify for a limited investigation; as a general case, long-chain polymers meet this criterion—we exclude, for the present purposes, two- or three-dimensional cross-linked systems. Even this restriction leaves a great deal of chemistry to consider, ranging from natural cellulose to synthetic polyethylene, from elastomers with their 'memory' of their unstretched state to DNA molecules with their store of genetic information.

This book has so far treated crystals as being built up from a large number of identical chemical motifs, arranged in a regular, repetitive order. Each motif may be an atom, a collection of ions, or the whole of, part of, or more than, one discrete molecule. In the crystallization process, the crystal grows by attaching further motifs to itself, taking them from the surrounding melt, solution, or vapour; the motifs are manoeuvred into the appropriate orientation under the influence of the forces that hold the lattice together; in the mobile phases, they are relatively free entities, with freedom of orientation. In other words, there is a sharp distinction between the two phases in crystallization; conversely, such crystals—provided they do not decompose—will show a sharp melting or sublimation point.

In contrast to this behaviour, glasses show no definite melting point: both organic and inorganic glasses continue to have extensive bonding networks throughout their bulk even when obviously molten. There is no very obvious distinction between the solid and liquid states. For instance, when molten soda glass is cooled, its viscosity progressively increases and the 'solid' that eventually results is really a very viscous supercooled liquid. Over a very long period of time, crystallization can set in and the glass then *devitrifies*.

In most long-chain polymers, an intermediate position is reached: because the bonding is extensive in one direction only, in contrast to that found in the glasses, some local ordering and crystallization is possible. The very long molecules can normally manage only this partial ordering because of their convoluted nature; they are also often irregular in their internal molecular stereochemistry. If, on the other hand, a perfectly-regular polymer backbone exists and the sample is annealed for long enough, good single crystals of polymers can be grown. Organic polymers therefore exhibit the full range of behaviour, from pure glasses to single crystals. This may even be seen within one sample, as in the elastomers, which tend to be amorphous when relaxed and ordered when stretched. Polyisobutene, for instance, gives characteristic-ally different diffraction patterns in these two states.

In such an elastomer, the resistance to stretching increases with extension, as for a metal spring; the mechanical deformation increasingly causes the polymer chains to reorder themselves into a more perfect arrangement; the more perfectly they fit together, the more strongly cohesive they are, and the greater the modulus of the elastomer. Many fibres behave similarly: nylon or polyester (terylene) are quite weak when first spun but if their fibres are stretched by several hundred percent, their strength increases to a similar degree, as ordering is imposed on the chains. Polyester films are biaxially stretched for the same reason. Since mechanical properties are so closely linked to internal order, X-ray diffraction becomes very valuable in the study of polymers. We shall consider four main fields of activity: degree of crystal-linity, the existence of preferred orientations, and the microstructure, which describes the sizes and distribution of the different regions or phases. The area of *microstructure* is much more akin to full crystal structure determination, in so far as it can be studied by diffraction methods, since it involves the details of the molecular structure.

Degree of crystallinity

As stated above, the tensile strength of polymer fibres depends on the degree of internal ordering, or crystallinity. Quantitative methods of measuring

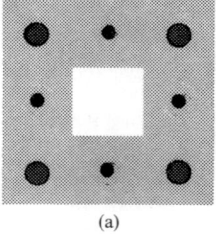
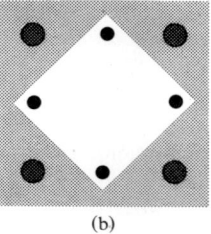

(a)   (b)

FIG. 5.6. (a) Defect concept; (b) perturbed packing concept.

this parameter are therefore of great practical importance. One point needs stressing: the definition of crystallinity is to some extent tied to the means used to measure it. The idea of distinct amorphous and crystalline zones within a polymer specimen is useful but actually rather arbitrary, yet it is extensively used. An alternative view considers local distortions or imperfections from an otherwise ordered lattice. The two approaches can be illustrated through a simple analogy, shown in Fig. 5.6. Part of a simple lattice, containing two kinds of atoms or ions, is shown; an atom or ion of one kind is accidentally missing from its proper site. We may think of this imperfection as a defect, as in (a), or as a packing error, as in (b). In real ionic crystals, the concentration of such imperfections is very low and it is therefore natural to think of them as defects. Their presence is manifested through conductivity and colour effects; were their concentration much higher, they might be detected by X-ray diffraction, through the presence of the lattice distortions.

The point of the analogy is that the bulk measurements, such as those of conductivity, will record a different proportion of affected crystal than the diffraction measurements. With polymers, one may use density comparisons, n.m.r. or i.r. spectra, or X-ray diffraction to specify the degree of crystallinity, obtaining a different answer in each case. The result hinges, of course, on which atoms appear to be in ordered chemical environments; this is differently defined for each approach.

In any X-ray diffraction experiment, a good part of the energy is scattered by the crystal as incoherent (i.e. non-Bragg) scattering. In single-crystal work, much trouble is taken either to exclude this from the detector or to allow for it with the aid of background-intensity measurements. X-ray measurements of the degree of crystallinity are performed by recording the relative intensities of the Bragg and non-Bragg scattering. Basically, the data used is that of Fig. 5.7, which combines two sets of measurements: the upper curve was recorded for a partially crystalline material, the lower curve for a completely glassy specimen of the same material. The difference in intensity as a function of θ is interpreted with reference to the form factors of the atoms in the material—the theory is too complex to be derived here but is given in detail

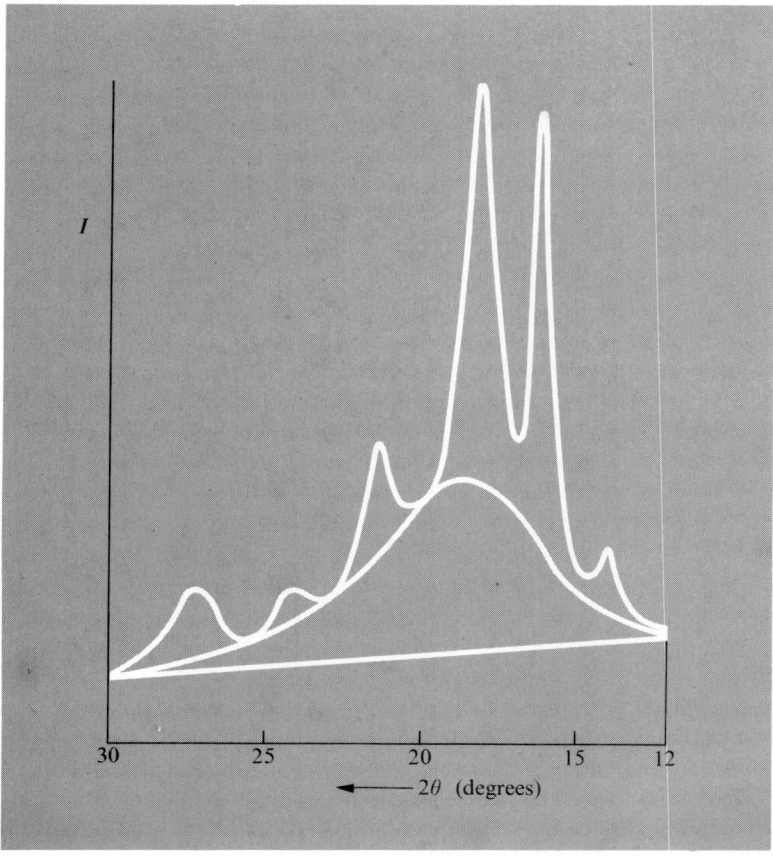

Fig. 5.7. Intensity versus Bragg angle for diffraction from isotactic (upper curve) and amorphous (lower curve) polystyrene.

in such textbooks as Alexander (1969). Data are normally collected by one of two common techniques.

In the first, Laue-type photographs are obtained from a thin film, as shown in Fig. 5.8 (shown here with two photographic plates installed to record forward- and back-reflections). If the film is wholly amorphous, diffraction effects can only arise between randomly-ordered atoms; these atoms are, however, kept within certain ranges of distance from each other by the existence of chemical bonds. Consequently, concentric diffuse rings will appear on the photographic plates, with 2θ corresponding to the median of each

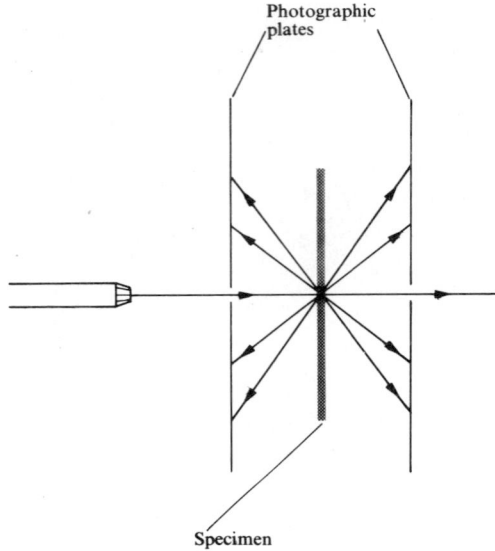

FIG. 5.8. Diffraction from a thin film.

group of bond lengths. The ring form arises in the same way as the sharp lines in the powder method, from cones of diffracted beams (see Fig. 5.2). If there is any one-dimensional ordering parallel to the incident beam, then the rings will become sharper; three-dimensional ordering produces intersecting cones about the ordering directions or axes, and the spots of the Laue photograph lie at the intersections. The film technique is useful because polymer chains in unstretched films often align themselves along the normal to the faces. In a stretched fibre, the orientation follows the fibre axis, and a technique similar to the rotation photograph of Chapter 3 is adopted. The fibre is stretched on a suitable frame, a collimated X-ray beam strikes it at 90° to the fibre axis, and reflections are recorded on a cylindrical film wrapped about that axis. No actual rotation need be used, since the crystalline regions are in random azimuthal orientation about the fibre axis. Periodicity along it is revealed by the appearance of layer lines in the photograph. As in the rotation photograph, the spots cannot immediately be indexed, except with respect to the mounting axis.

Note that axis-length in a fibre has somewhat different connotations from that in a normal three-dimensional crystal. In ordinary lattices, the unit cell contains chemical motifs, often comprising one or more molecules; the lattice translations relate exactly similar unit cells to each other. In a polymer chain,

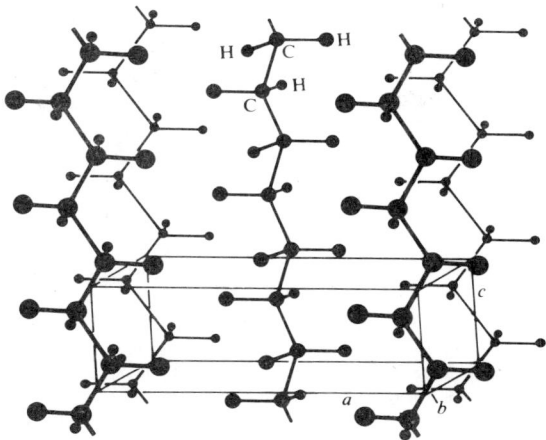

FIG. 5.9. Perspective view of the orthorhombic crystal structure in polyethylene. Each unit cell contains two $-CH_2-$ repeat elements.

the situation is different, since the periodicity arises *within* the molecule. Unit cells containing whole molecules would be quite impractically long; furthermore, chain-lengths vary appreciably. The unit cell in the polymer is therefore tied to stereochemical repetitions along the chain; these are considered on p. 64. Figure 5.9 shows the unit cell in a polymer single-crystal.

Orientation

Polymer chains, then, represent a measure of one-dimensional order in themselves. When they are aligned together, usually by stretching the specimen, overall one-dimensional order appears and mechanical properties change. Polymer scientists therefore wish to know to what extent the chains have become aligned in a particular material, with particular reference to fibres. If an X-ray beam is directed along the fibre axis, anything from diffuse rings to a full-scale Laue photograph may result, according to the degree of crystallinity, which can range from that of a glass to that of a single crystal. Usually, an intermediate situation is found, so that the Laue spots are more or less smeared out, as a function of the misalignment of the crystalline regions. This is a three-dimensional misalignment which is therefore not easy to represent. The best technique is to use a four-circle diffractometer to scan reciprocal space; a number of reflections on the surface of a sphere about the origin of the reciprocal lattice are examined and the intensity variation about their nominal positions is recorded by suitable sampling. The variations can then be plotted as a contour map on a globe, in relation to the principal axes.

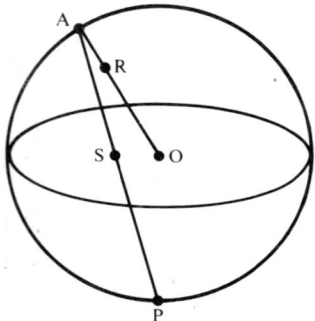

FIG. 5.10. Spherical and stereographic projections. R is a point in the reciprocal lattice, with origin at the centre of the sphere, O. OR is projected to cut the sphere at A. The line AP runs to the south pole, P, and cuts the equatorial plane at S. The map of points like S on the equatorial plane forms a two-dimensional projection of the lattice of points R.

This map is best represented in two dimensions by a *pole figure*, whose construction and interpretation is explained by Alexander (1969). It is closely related to the *stereographic projection* of Fig. 5.10, which is valuable in representing the symmetry elements of the three-dimensional crystallographic point groups.

Micro- and macro-structure in polymers

The microstructure of a polymer comprises the arrangements of the chains, both internal and in their packing together. Although the packing and crystallinity can be profoundly affected by the way in which the finished polymer is handled, thermally and mechanically, they must also be controlled by the stereochemistry of the chains themselves. As in the case of the detection of periodicity from diffraction patterns, what matters is the nature of the repeat unit in the chain. Certain useful classifications have been generally adopted. Thus a *tactic* polymer must contain monomers joined together in a regular head-to-tail order, without cross-linking or branching. Only such a polymer can form perfect crystals; *atactic* polymers are those that cannot meet these conditions. There are two further subdivisions of tactic polymers, according to the way in which substituents may appear along the chain; the pattern is either one of exact repetition (ABABAB pattern, *isotactic*, Fig. 5.11(a)) or of alternating repetition (ABBAABBA pattern, *syndiotactic*, Fig. 5.11(b)). The figures also show the repeat distances associated with the two patterns.

Catalysts with a very stereospecific mode of action are needed to ensure tacticity and they produce a distinctly different product from non-stereo-specific catalysts. Different catalysts can produce quite different results, even

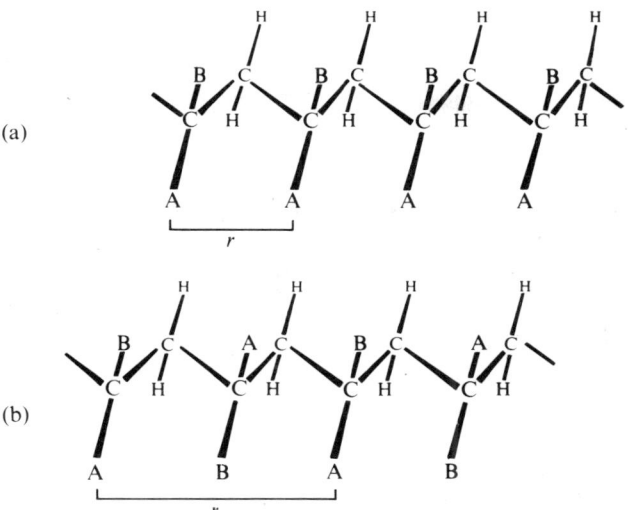

FIG. 5.11. (a) Isotactic and (b) syndiotactic polymers. r indicates the crystallographic repeat unit.

when tacticity as such in not involved. Polyethylene, for example, is available in two quite different forms, high-density polyethylene (HDPE) and low-density polyethylene (LDPE). The former is made with Ziegler–Natta catalysts, generally based on a reaction product of $TiCl_3$ and an aluminium trialkyl; the active catalyst is an organometallic species containing a titanium–carbon σ-bond, into which successive ethylene molecules insert.

LDPE, in contrast, is made by free-radical polymerization, with the front of the growing chain attaching successive monomer molecules. As the names imply, HDPE is appreciably denser than LDPE; it is also tougher and has a higher softening temperature. The difference in physical properties presumably reflects differences in microstructure.

Perhaps the most significant single structure of a linear polymer to be resolved with the aid of X-ray diffraction data is that of DNA. The distinctive helical configuration of the two strands of the DNA molecule was deduced from the evidence of periodicity in the diffraction photographs of the crystalline material. Watson's highly readable account of the speculations that led to the correct structural model is recommended (Watson, 1968).

So far, all our thoughts have been concentrated on diffraction effects produced by interference between scattering centres separated by distances ranging from 100 pm (the approximate diameter of an atom) to 1000 pm (the separation of planes below which Bragg angles become inconveniently small).

In other words, conventional X-ray diffraction, using 2θ angles from perhaps 5° to 180°, deals with atomic and molecular systems. In principle, scattering from larger systems ought to produce interference effects from the same wavelengths of incident radiation, provided observations are made at sufficiently small values of θ. A *low-angle diffraction* apparatus for this purpose can be constructed. The essential features are : the use of highly mono-chromatic X-rays ; very fine collimation ; a great distance between specimen and photographic plate; evacuation to remove air as a source of extraneous scattering. With such an apparatus values of 2θ down to 15′ of arc can be recorded. Two kinds of intensity distribution are recorded : from *diffuse* and *discrete* scattering. In either case, scattering regions with dimensions from 3000 to 100 000 pm can be examined.

Diffuse scattering arises by interference between points *within* a relatively homogeneous region—just as the form factor curve is the result of interference between different points in the electron shells of an atom. The variation of diffuse-scattering intensity with angle can therefore be used to measure the size and shapes of independently scattering but similar regions. The main applications are to dilute suspensions of colloids, long-chain polymer mole-cules in solution, and globular proteins. Discrete scattering becomes important when the scattering regions come sufficiently close together for interference to occur between them. The intensity distribution as a function of $\sin\theta/\lambda$ may then be applied to the study of polymer morphology and macrostructure ; it is particularly useful for following the folding of chains, which is a prominent feature of many polymers.

Other applications of X-ray diffraction

The specialized techniques used with polymers have a number of applica-tions to other materials : a few illustrative examples will be quoted, covering both wide- and small-angle diffraction. For example, the tendency of polymer chains to lie along the normal to the plane faces of unstretched films is mimicked by the orientation of grains in certain metals and alloys. When metals with the cubic f.c.c. lattice are cast into thin sheets, the crystal grains tend to orient their 100 axis perpendicularly to the casting table, while the orientations of the other two axes about this vertical remain random. Such a structure is called a *texture*. Again, polymer chains tend to align in the direction of mechanical extension of the bulk article, while the grains in a cold-rolled metal follow the direction of deformation, tending to lie perpendicularly to the nip of the rollers and within the plane of the sheet produced. Just as in polymers, so in alloys the orientation of these crystalline regions profoundly affects the mechanical properties. The degree of orientation can be recorded by dif-fraction techniques similar to those used for polymers. Small-angle scattering can tell us the size and shape of the grains themselves.

The lower curve of Fig. 5.7 is not a smooth decrease of intensity with rising θ. Instead, it shows a definite maximum. Even an amorphous polymer, therefore, shows a certain diffraction effect. For polystyrene, evidently, this comes from the probability of finding atoms separated by normal bonding distances, spread over a certain statistical range. Even a completely random structure, therefore, will produce diffraction rings, although these will be diffuse, and we may hope to extract some information from these rings. The three random states of matter are gas, liquid, and glassy. The first is distinguished by the relatively high dilution of chemical species within it; this greatly simplifies the diffraction pattern and increases its value, which will be discussed separately in Chapter 7, p. 81. The structures of glasses have been extensively examined, particularly in the hope of detecting incipient local order. The modern view of glasses is that they are by no means homogeneously random; the presence of regions of different kinds seems to be more marked in two-component than in one-component glasses, presumably owing to limitations on the mutual solubility of components. Small-angle X-ray diffraction has been used to detect these regions and wide-angle diffraction to establish the distribution of covalent bond-lengths in glasses. Urnes (1960) has published an account of this work.

As glasses contain stable covalent bonds, the presence of some limited order is not too surprising—they do, after all, tend to crystallize and devitrify in the long run. Molecular liquids, on the other hand, would seem to contain molecules that constantly jostle, rotate, and exchange places, so that no order is retained beyond a fleeting instant. This model implies relatively weak and non-directional intermolecular forces; such a model liquid would also form ideal solutions and be smoothly miscible with other such liquids. We do in fact know that liquid mixtures and solutions can show considerable departures from ideal behaviour; such behaviour is attributed to the action of specific forces tending to draw like to like. Brady (1971) has published a fascinating account of small-angle X-ray diffraction work that provides direct evidence of these effects. As the lower consolute temperature of liquid pairs is approached, the *persistence length* of order within the single phase grows very quickly. The persistence length is the distance from any given centre over which structural similarity is maintained. In fluorocarbon–hydrocarbon mixtures, persistence lengths of up to 4000 pm are observed several degrees above the critical temperature at which two phases separate—in other words, the macroscopic phase separation is preceded by a rapid growth in the banding together of like molecules with like. Another study involved *liquid crystals*, in which elongated molecules such as *p*-azoxyanisole tend to line up together in the liquid state below a certain critical temperature. Here, preliminary ordering above the critical temperature was found to extend over ranges of up to 185 000 pm.

6. Neutron diffraction

The diffraction of thermal neutrons

THE de Broglie relationship endows every particle with an associated wavelength which is a function of its mass and velocity. A neutron, which has a mass of 1.66×10^{-27} kg, will acquire an associated wavelength of 100 pm when its velocity is about 3900 m sec^{-1}. For comparison the mean velocity of a hydrogen molecule in the gas at 25°C is 1768 m sec^{-1}. The associated wavelength of our neutron is between the characteristic Mo-K$_\alpha$ (71·07 pm) and Cu-K$_\alpha$ (154·18 pm) lines. Neutrons with such velocities are called *thermal neutrons* and may be obtained at high flux densities by *moderating* the velocities of the fast neutrons formed in nuclear reactors through repeated collisions with the atoms of a moderator such as graphite. The process of deceleration by random collisions produces a virtually Gaussian distribution of velocities about a mean value and a collimated beam of such neutrons can be used to produce Laue photographs of single crystals. Monochromatic radiation is generally preferred for diffraction experiments and can be approximated by selecting only a small part of the velocity distribution, either by time-of-flight resolution with successive rotating segments, or with a monochromator. Since there is no characteristic strong peak at any velocity, the narrower the slice of the velocity distribution the lower its intensity, so that a compromise has to be reached between purity and strength in the primary beam. A further disadvantage is the very weak response of photographic films to neutrons incident upon them. Powder or single-crystal diffractometers have to be used and must be massively constructed to support the heavy shielding that protects the detector from stray neutrons.

The main problem, of course, is to find a sufficiently intense source of thermal neutrons. Although some work is in hand on the use of proton accelerators to bombard targets and produce intense bursts of 'white' radiation, with which several reflection intensities are measured at once with several detectors, a nuclear reactor is usually needed. Not all reactors can produce very intense beams and rather large crystals (up to $5 \times 5 \times 5$ mm) sometimes have to be used to produce adequately strong reflections. Counting times must also be longer than with X-rays to produce the same statistical reliability in the intensity measurements. Historically, these factors have severely limited access to neutron diffraction. Why, then, does anyone bother with it? The answer is that neutron diffraction measurements not only complement X-ray data but also yield unique information on the magnetic structures of solids, as we shall see below.

Elastically scattered neutrons

X-rays are scattered by the orbital electrons surrounding the atomic nucleus. To a first approximation, the spherical overall distribution of the electrons about the nucleus of a free atom or ion is not disturbed by chemical bonding to other atoms. Even in compounds, therefore, the nucleus sits at the centre of mass of the electron cloud, so that the position of the nucleus can be fairly accurately deduced from X-ray diffraction data. Covalent bonding naturally tends to upset the validity of the approximation but the discrepancy is only serious for the lightest elements, with fewer core electrons in relation to valence electrons. Neutrons are directly scattered by the atomic nuclei and a structure analysis based on their diffraction ought to give much the same results as X-ray work. Hydrogen atoms, however, appear to be in significantly different positions because the electron density is drawn into the covalent-bond axis, and the electron-cloud displaced relative to the nucleus. The analysis of neutron diffraction patterns follows the path of Patterson, Fourier, and least-squares refinement techniques described in Chapter 2.

Significant differences between X-ray and neutron diffraction patterns are the result of the different scattering mechanisms, of which there are three for neutrons. The simplest is *elastic scattering* in which the neutrons bounce off the nuclei without any particular change of intensity or velocity with angle of deflection. Interactions with the orbital electrons are taken to be negligible. As atomic nuclei have radii of about 10^{-2} pm, 100 pm wavelength neutrons show no interference effects between different parts of the nucleus.

Whereas the X-ray form factor, f_x, varies with $\sin\theta/\lambda$, the neutron form factor, f_n, is constant. Again, the scattering amplitude for atoms increases with atomic number when X-rays are used but only undergoes minor fluctuations with neutrons, as shown in Fig. 6.1. The negative values for certain nuclei

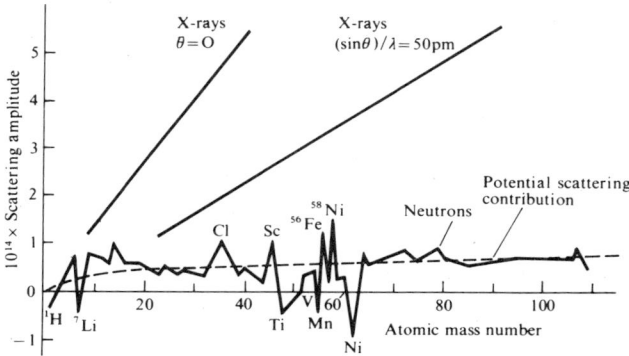

FIG. 6.1. Scattering amplitudes for neutrons and X-rays as a function of atomic mass number.

indicate that a 180° phase-change occurs on collision with these. The salient point of the figure is that f_n for lighter atoms such as H, C, O, or N are not very different from the values for quite heavy elements such as Fe, Ni, or Pd. This makes possible much more accurate location of light-atom positions in structures containing such heavy elements than is possible with X-rays. The improvement is naturally most marked for hydrogen atoms, especially as neutron diffraction reveals the true position of their nuclei. Hydrogen atoms play a unique role in chemistry, and we shall consider two topics involving them.

Hydrogen bonding is a weak interaction, whereby hydrogen atoms covalently bonded to electronegative elements such as nitrogen, oxygen, or chlorine appear to interact with other such atoms to which they cannot formally bond by any covalent mechanism. Despite the weakness of individual hydrogen bonds, they can act cumulatively and thus influence the structures of solids to a considerable extent. Ice at 0°C, for example, is less dense than water at the same temperature; this implies a kind of ordering that increases the free volume within the crystal on solidification.

The X-ray structure of the low-temperature form of ice appears to be a loose packing of the oxygen atoms into tetrahedral 4:4 coordination, with the hydrogen atoms unlocated. The i.r. spectrum suggests a weakening of the O—H bond relative to that in water vapour. Heat-capacity measurements show that the entropy of the crystal does not tend towards zero at the absolute zero of temperature; instead, there seems to be disordering between two possible states, leading to residual entropy.

The neutron-diffraction study of Peterson and Levy (1957) on D_2O at $-50°C$ supplied a structural explanation. The observed coordination of an oxygen atom is depicted in Fig. 6.2, where the notation $\frac{1}{2}D$ represents a scattering centre with the properties of a deuterium atom but only half its power. This *fractional occupancy* of a site is a commonly observed effect in crystallography and is the result of the random presence or absence of an atom at equivalent sites in different unit cells throughout the crystal. Here we have sites within bonding range of one or other oxygen atom alternately filled

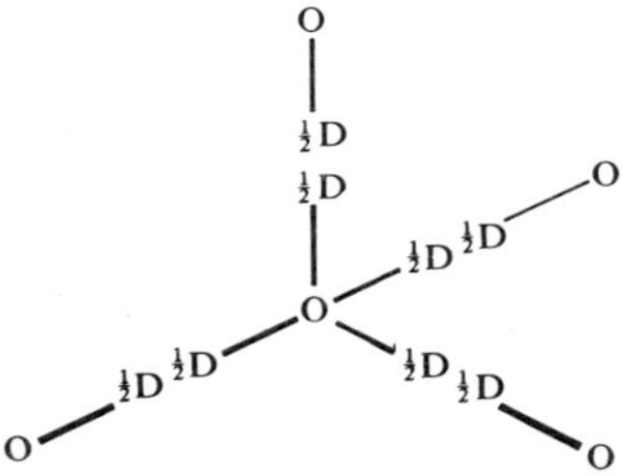

FIG. 6.2. The observed coordination of oxygen in solid D_2O at $-50°C$.

$$\tfrac{1}{2}(O\text{---}D\text{-----}O + O\text{-----}D\text{---}O) \longrightarrow O\text{-----}\tfrac{1}{2}D\tfrac{1}{2} D\text{-----}O$$

FIG 6.3. Disordering of hydrogen bonding in solid D_2O.

and empty. The explanation is shown in Fig. 6.3, in which the two possibilities combine.

Two qualifications should be noted: the thermal vibration parameters of the two $\tfrac{1}{2}$D's are relatively normal; there is a significant lengthening of the O—D bond. These observations indicate that the deuterium atom is not simply undergoing wild oscillations between the two $\tfrac{1}{2}$D positions, but that it is affected by the proximity of the non-bonded oxygen. In fact, there is a hydrogen-bonding interaction which knits the water molecules together into an ice crystal. The network of hydrogen bonds is not completely ordered, as deuterium atoms could switch from one $\tfrac{1}{2}$D position to the other by concerted movement throughout the crystal; hence the frozen-in excess entropy. Deuterium oxide was used in this study because it produces less diffuse-scattering background than normal water.

Hydrogen bonds of this kind are very common, particularly in controlling the detailed configurations of large biological molecules. The matching of the strands in DNA, for example, is controlled by the hydrogen-bonding inter-actions between the purine and pyrimidine base pairs, shown in Fig. 6.4. Not

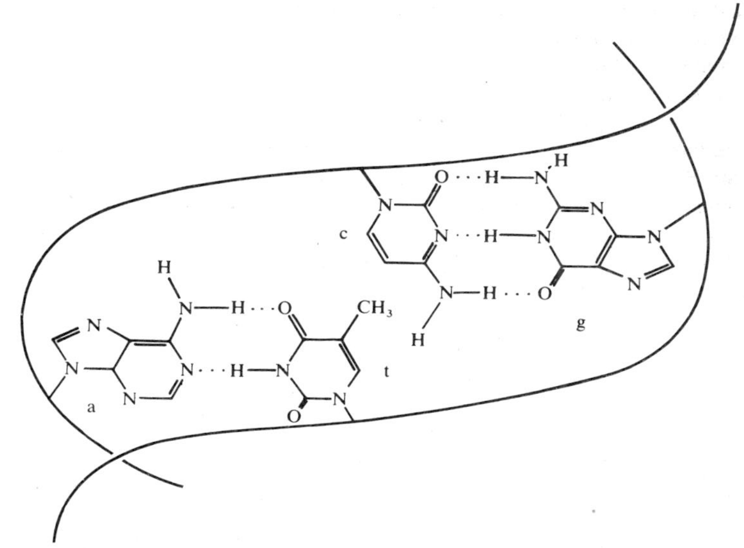

FIG 6.4. Base-pairing in DNA. The continuous lines represent the sugar–phosphate chains. a = adenine, c = cytosine, g = guanine, t = tyrosine.

all hydrogen bonds are asymmetric—a notable exception is the linear HF_2^- anion in KHF_2, in which the hydrogen atom is symmetrically located between the fluorines, as was proved by neutron diffraction.

Apart from hydrogen bonding, single-crystal neutron diffraction has been used to find the positions of hydrogen atoms in organic compounds, inorganic hydroxides, hydrides, and ammonium salts, and fluorine atoms in the xenon and uranium fluorides. In the organic compounds, the 'neutron-diffraction determined' C—H bond-length is approximately 108 pm, as opposed to an apparent 80 pm from X-ray work (cf. p. 69). For a general review of structural analysis by neutron diffraction, a review article by Bacon (1964) may be consulted.

As our second illustration of the complementarity of X-ray and elastic neutron scattering, we shall briefly consider transition-metal hydride complexes, a subject not covered by Bacon. Their structures originally provoked a minor controversy, now resolved. When complexes containing a hydrogen atom covalently bonded directly to the central metal atom were first studied, there was a good deal of confusion. Many held that the hydrogen—formally represented today as a hydride ligand, H^-—was either abnormally tightly bonded to the metal atom or even actually buried within the latter's electron orbitals. Chemical and spectroscopic data were interpreted to support the contention that the ligand–metal bond was abnormally short and that the ligand occupied no definite position in the coordination environment of the metal atom. This theory is now totally discredited but the history of one such complex will show how far ingenious (but wrong) speculation can go in the absence of definite structural information.

Pentacarbonylmanganese hydride, $HMn(CO)_5$, was thought by Hieber—the pioneer of transition-metal hydride-complex chemistry—to contain the Mn–H 'pseudo-iron atom' (Hieber, 1958). Broad-line n.m.r. data from solid samples appeared to give an Mn—H bond length of some 143 pm—well below the sum of the normal covalent radii, which is usually a tolerable guide to single-bond lengths. Interestingly enough, an electron-diffraction study was made to confirm this result.

X-ray diffraction, however, clearly suggested that the hydride ligand, although not directly 'seen', did occupy a coordination position (Ibers, 1965). Thus the metal-carbonyl moiety of α-$HMn(CO)_5$ adopted the square-pyramid form shown in Fig. 6.5. The hydride ligand has been added to the drawing in its presumed position at the sixth apex of an octahedron about the manganese atom.

β-$HMn(CO)_5$ (which is merely a different crystalline form, otherwise chemically similar) was subjected to parallel X-ray and neutron diffraction studies by La Placa, Hamilton, Ibers, and Davison (1965), which clinched the argument. The X-ray data, taken at $-75°C$, confirms the geometry of Fig. 5.5, without showing the hydride ligand; the neutron study at $-105°C$ revealed it

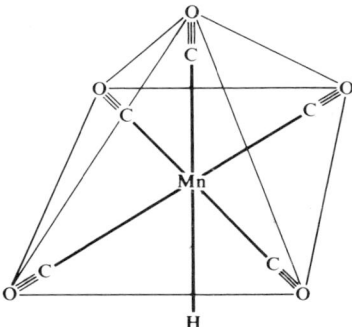

FIG. 6.5. The structure of $HMn(CO)_5$.

in the expected position, $160·1 \pm 1·6$ pm from the manganese, a distance close to the sum of the covalent radii.

Today, the hydride ligand is so well accepted that its position is often inferred from the existence of gaps in the coordination environment. Thus the first triply-bridging hydride ligand was postulated in $H_2Ru_6(CO)_{18}$ (Churchill and Wormald, 1971). The molecule is shown in Fig. 6.6. Note the way in which the carbonyl ligands are bent back from the two triangular faces of the octahedron parallel to the plane of the paper. The invisible hydrides are taken to lie above these faces, symmetrically bonded to the three ruthenium atoms in each face.

Finally, the ultimate proof that hydride ligands are really such is surely found in the neutron diffraction structure of K_2ReH_9, which contains the

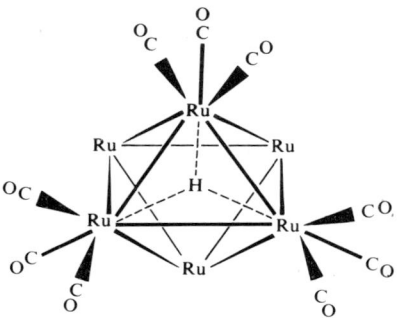

FIG. 6.6. The structure of $H_2Ru_6(CO)_{18}$. The more-distant Ru_3 triangle is linked to the closer Ru_3 triangle by a centre of symmetry and its nine carbonyl and one hydride ligands are omitted for clarity.

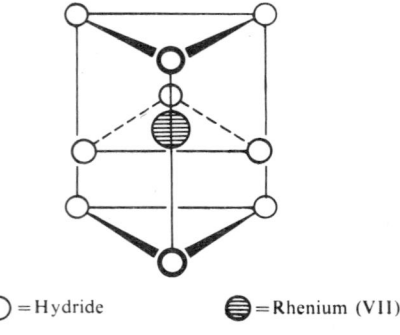

$\bigcirc$ = Hydride $\ominus$ = Rhenium (VII)

FIG. 6.7. The $[Re^{VII}H_9]^{2-}$ dianion.

ReH_9^{2-} dianion of Fig. 6.7. This is also a good example of the unusual nine-coordination, in the shape of a tricapped trigonal prism.

Magnetic scattering of neutrons

Elastic scattering of neutrons is not affected by the presence of the orbital electrons, according to our previous assumptions. This, on closer examination, proves not to be universally valid. The neutron is more than a small billiard ball that bounces off nuclei: it has a magnetic moment. So has any atom containing one or more electrons with unpaired spins; the magnetic moment of such an atom is a combination of moments contributed by the electron spins and orbital angular momenta. The atomic and neutron magnetic moments will interact according to a force law and, because the atoms are for all practical purposes rigidly held in their lattice, the neutrons will undergo an additional *magnetic scattering*. A *magnetic form factor*, f_m, has therefore to be added to f_n. A complication arises here: the magnetic scattering results from an appreciably large electron distribution, so that f_m is a function of θ and λ (p. 14). It is naturally also a function of the angle between the two magnetic moments.

If either the magnetic moments of the incident neutrons or those of the magnetic atoms in the lattice are randomly oriented, then the magnetic scattering merely appears as an additional incoherent component added onto the background to the elastic scattering but distinguishable from it through its angular dependence of intensity. If, however, a polarized neutron beam is used, in which the magnetic moments are aligned in one direction, much information can be obtained on those crystals in which the atomic magnetic moments tend to line up together. The principles and results of this kind of magnetic neutron diffraction have been discussed by Bacon (1966b); we shall review the information obtained on the four principal conditions of cooperative magnetization—paramagnetism, ferromagnetism, antiferromagnetism, and ferrimagnetism.

These four states are identified by the characteristic shapes of their magnetic-susceptibility (χ) versus absolute-temperature (T) curves, which are diagrammatically represented in Fig. 6.8. Paramagnetic solids are so defined here as to exclude the special cases of temperature-independent (Van Vleck) and free-electron (Pauli) paramagnetism. χ is then seen to vary as T^{-1} (Curie Law) or as $(T-\theta)^{-1}$ (Curie–Weiss Law). The susceptibility is a measure of the ability of the applied magnetic field to force individual magnetic dipoles within the specimen into alignment. Thermal motion opposes the alignment. Each dipole consists of an atom with one or more unpaired electrons; the Curie Law governs specimens in which these magnetic atoms are too far separated to interact appreciably, while the Curie–Weiss Law describes cases in which there is significant interaction. These are often called *magnetically-dilute* or *non-dilute* systems. A typical paramagnetic compound is potassium hexacyanoferrate(III), $K_3Fe(CN)_6$; the basic magnetic unit is the $[Fe(CN)_6]^{3-}$ ion, with one unpaired electron in a 3d orbital of the iron atom. The randomly-oriented magnetic dipoles in an unmagnetized sample of $K_3Fe(CN)_6$ will add a diffuse scattering component to the background in diffraction experiments but no more.

Ferromagnetic substances show a very high magnetic susceptibility at low temperatures but this falls off rapidly as the sample is warmed. Above the Curie temperature T_C, normal paramagnetic behaviour is resumed—see Fig. 6.8(b). The high susceptibility at low temperatures is the result of strong cooperative interactions between the magnetic dipoles, which allow a very high degree of magnetic ordering in an applied field. If a sample of a ferromagnetic material, with dipoles aligned in a magnetic field, is exposed to a beam of polarized neutrons, true magnetic diffraction can be observed. As stated above, the infensity of scattering is a function of the angle between the dipoles in the sample and those in the beam, and is in fact controlled by eqn (6.1), where M is the scattering amplitude for angle α and M is that when α is a right angle. Conversely,

$$M_\alpha = M_{90} \sin \alpha. \tag{6.1}$$

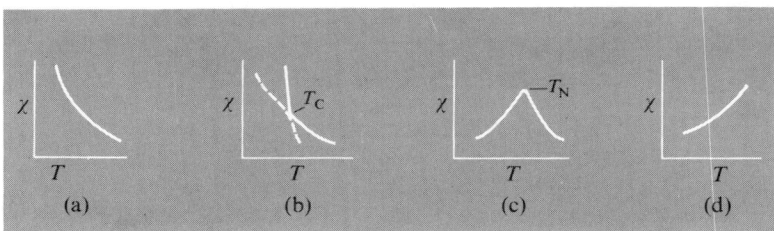

FIG. 6.8. The temperature dependence of magnetic susceptibility for (a) paramagnetism, (b) ferromagnetism, (c) antiferromagnetism, (d) ferrimagnetism.

Bragg planes that contain the direction of magnetization make $\alpha \neq 0°$ in all circumstances, whatever the value of θ, and therefore tend to polarize a neutron beam reflected from them. Common ferro-magnetic materials studied by neutron diffraction include b.c.c. iron and f.c.c. cobalt. It is even possible to distinguish between the various magnetic contributions made by the different atoms in ferromagnetic alloys as a function or their environment.

Antiferromagnetic materials contain two independent sets of magnetic dipoles, which, if undisturbed, tend to align themselves in mutual opposition. The opposed moments cancel, producing a low-temperature susceptibility more characteristic of diamagnetism than paramagnetism. As the temperature is raised, those moments in the 'wrong' direction with respect to the applied field increasingly tend to break loose and conform to it, causing a rapid increase in susceptibility. Above the Néel temperature, T_N, however, the general breakdown in cooperative ordering overwhelms this initial trend and paramagnetic behaviour is resumed, as seen in Fig. 6.8c.

The original—and, as it turns out, simplistic—interpretation of the antiferromagnetic behaviour of metallic chromium is depicted in Fig. 6.9, as deduced from the early neutron-diffraction experiments. Chromium has the b.c.c. structure, which consists of two interpenetrating cubic lattices, whose constituent atoms are represented by open circles in the figure. The superimposition of the two lattices produces the set of systematic absences governed by eqn (6.2), characteristic of body-centring.

$$h+k+l = 2n+1 \tag{6.2}$$

X-ray or elastic neutron diffraction cannot distinguish between the atoms of the two different lattices and the absences are therefore observed in their diffraction patterns. When oriented magnetic moments are assigned according to the arrows the two lattices become quite distinct in magnetic scattering, so

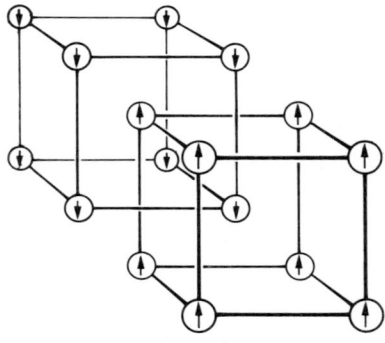

FIG. 6.9. A simplified representation of the magnetic structure of b.c.c. chromium.

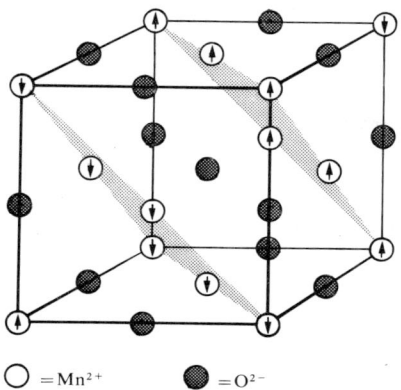

$\bigcirc$ = Mn^{2+} ● = O^{2-}

FIG. 6.10. The crystal and magnetic structure of $Mn^{II}O$.

that normally absent reflections such as (100), (010), or (111) become observable. This model gives a good, although far from perfect, fit with the experimental data—the exact situation is actually much more complicated.

In other cases the *magnetic unit cell* is differently related to the normal unit cell. Whereas in b.c.c. chromium it is merely the primitive, non-centred cubic cell, in manganese(II) oxide, MnO, it is a multiple of that cell. In such a case, a *magnetic superlattice* exists, so that magnetic reflections are observed, with fractional indices based on the normal unit cell. The effect is analogous to that produced by superlattices in substitutional solid solutions on elastic diffraction patterns. The structure of MnO, shown in Fig. 6.10, is essentially like that of rock salt, with interpenetrating f.c.c. lattices of Mn^{2+} and O^{2-} ions. The basic magnetic unit is the Mn^{2+} ion, with five unpaired electron-spins in its 3d orbitals. These magnetic moments are aligned in groups corresponding to the f.c.c. close-packing planes of the Mn^{2+}-lattice, with alternate planes showing opposite orientations. The planes have been shaded in the figure. Note, however, that they represent only one of the three possible sets of the general $\langle 111 \rangle$ form, according to the choice of axis labels in the cubic unit cell.

Within any one *magnetic domain* of the crystal, therefore, the planes of similarly oriented Mn^{2+} ions are all parallel but not so between different domains. Bulk susceptibility measurements cannot reveal this, but for magnetic diffraction work to succeed a magnetic single crystal that is one single domain must be obtained. The chromium and manganese(II) oxide structures illustrated are quite simple; neutron-diffraction studies have revealed far more complex arrangements in other systems, such as those shown in Fig. 6.11. Helical arrangements of magnetic moments, for example, appear to be particularly common in the rare-earth metals and their alloys.

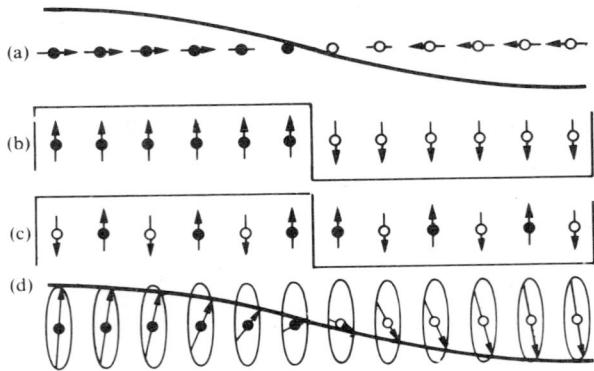

FIG. 6.11. Some kinds of long-wavelength modulation in antiferromagnetic crystals. The arrows represent the direction and degree of orientation of magnetic moments in successive crystal planes: (a) sinusoidal, (b) square wave, (c) out-of-step domains, (d) helical.

In ferrimagnetic structures the opposing magnetic moments are again disposed on interpenetrating lattices. In contrast to those in antiferromagnetic materials, they do not exactly cancel out at the absolute zero of temperature because the distributions are unevenly balanced. The situation arises when there are two different kinds of paramagnetic ion, or when the same ion appears in different environments. The commonest class of such materials are the *spinels*. These have the general formula A_2BX_4 and the compound after which the class is named is Al_2MgO_4. In this, there is an f.c.c. lattice of O^{2-} ions, providing one octahedral and two tetrahedral interstices per anion. Half the octahedral sites are occupied by Mg^{2+} ions and one-eighth of the tetrahedral sites by Al^{3+} ions. Spinel itself is diamagnetic, since these cations contain no unpaired electron-spins.

The *inverse spinels*, such as Fe_2MgO_4, have a structure which is better represented as $(AB)AX_4$—in this particular example, the Fe^{3+} ions lodge in one-quarter of the octahedral and one-eighth of the tetrahedral sites; the Mg^{2+} ions occupy another quarter of the octahedral sites. Fe_2MgO_4, or ferrite, shows ferrimagnetism. The ferrites are very valuable for making the cores of radio-frequency coils and transformers, on account of their high magnetic susceptibility coupled with high electrical resistance, which minimizes losses from eddy-currents. Neutron-diffraction experiments have proved that the Fe^{3+} ions, with their five unpaired d-electrons, have their magnetic moments differently oriented according to the kind of sites they occupy. Néel's original hypothesis on the nature of ferrimagnetic materials has thus been directly confirmed. Other ferrimagnetic spinels are found in wide variety, based on the divalent cations Mn^{2+}, Fe^{2+}, Ni^{2+}, Cu^{2+}, Zn^{2+}, and Mg^{2+}, the trivalent cations Mn^{3+}, Fe^{3+}, Co^{3+}, Al^{3+}, and Ga^{3+}, and the anions O^{2-}, S^{2-}, and Se^{2-}.

An important aspect of cooperative magnetic behaviour is the mechanism whereby magnetic moments are coupled between non-neighbouring atoms. This phenomenon of *superexchange* involves a spin correlation transmitted through the electronic orbitals of intervening anions, themselves diamagnetic. The term was coined by reference to the exchange forces which correlate the spins of different electrons in the orbitals of atoms and molecules, producing the rule of maximum multiplicity. The occurrence of superexchange naturally depends on the degree of overlap between atomic orbitals, which in turn is controlled by the crystal structure. Neutron diffraction can thus both provide data on the magnetic structure and indicate the crystal parameters to which it is sensitive. The compound manganese(II) sulphide has been discussed by Bacon (1966b): there are three polymorphic forms, each with different degrees of superexchange and therefore different magnetic structures.

Inelastic neutron scattering

A third and more specialized aspect of neutron scattering involves energy exchanges between the lattice and the incident beam. The internal vibrations of crystals are quantized, with discrete energy-levels resembling standing waves in a stretched string or in the particle-in-a-box model of elementary quantum mechanics. The quantized units of vibrational energy, or *phonons*, can be exchanged with the neutron beam, leading to detectable energy-shifts in it, which yield information on the vibrational modes of the crystal. The best analogy is to the frequency shifts in infrared radiation produced in Raman spectroscopy by energy exchange with the vibrational levels of individual molecules. Although not a diffraction technique, inelastic neutron scattering yields valuable information, complementary to a knowledge of the crystal structure.

PROBLEMS

6.1. Which magnetic superlattice reflections would you expect to observe for the MnO structure in Fig. 6.10?

6.2. Why are ferromagnetism, antiferromagnetism, and ferrimagnetism found only in the elements of the transition period, the lanthanide and actinide series, and their compounds?

6.3. Solid matter absorbs X-rays far more strongly than it does thermal neutrons. Yet the intensity of Bragg reflections is generally taken to be proportional to F_{hkl}^2 for X-rays but to F_{hkl} for neutrons in diffraction work. Why?

7. Electron diffraction

Introduction

THE principles of interference for beams of electrons are the same as for neutrons but the charged nature of electrons makes them substantially different for use in diffraction experiments. Electrons can be accelerated through known potential differences to an accurate velocity; they may also be focussed, both before and after diffraction. We shall return to these practical advantages later. On these grounds, the charged nature of the electron should render it ideal for diffraction experiments; however, it also introduces both advantages and penalties in the scattering process. Electrons are scattered from atoms and molecules with a very different amplitude, compared to neutrons or X-rays; electron diffraction is therefore subject to different but complementary practical restrictions. This account of it will focus on those aspects that primarily interest chemists.

Electrons are chiefly scattered by interaction with the electric field of the atomic nucleus. This creates a potential with a much less sharp dependence on distance from the nucleus than for neutrons; interaction in fact becomes significant further out for electrons than even for X-rays. Consequently, interference effects are important—the atom that an electron 'sees' is very much larger than that for neutrons and slightly larger than that for X-rays. The electron form-factor, f_e, has a pronounced dependence on $(\sin \theta/\lambda)$, just as does f_x, but differs from the latter in magnitude: whereas f_x is usually of the order of 10^{-2} pm, f_e may reach 10^2 pm, and is very dependent on λ. In most instances, therefore, a high proportion of the incident electrons are scattered as opposed to a very small proportion of X-ray intensity. The kinematic theory of X-ray scattering must therefore be replaced by a dynamical theory for electrons. As a rough guide, there is low fractional scattering of very energetic electrons (several hundred keV), or of moderately energetic electrons (10–100 keV) from specimens of low atomic number; for most diffraction experiments with electrons below 100 keV, and certainly for all below 10 keV, the dynamical treatment is essential. For a more formal discussion of electron-scattering by atoms, a review article by Vainshtein (1964) may be consulted.

The corollary of the strong scattering is low penetration of the primary electron beam into specimens. Note that this is a power-loss largely by coherent diffraction, as opposed to the absorption loss of X-ray power by fluorescence and other incoherent processes that contribute to general background. The very different balance between penetration and coherent-scattering power has caused the development of electron diffraction to follow paths different from those of the X-ray and neutron techniques. Electron diffraction is mainly used with very thin specimens, such as films or surface layers, or with samples of dilute scattering power, notably gases and vapours. The ready availability of electron beams of different energy is a great experimental help; we shall break the field down into diffraction from vapours and gases, *high-energy electron diffraction* (HEED) from solids, and *low-energy electron diffraction* (LEED) from surfaces.

Electron diffraction from gases and vapours

The Wierl type of apparatus is shown in Fig. 7.1. A fine beam of electrons of uniform velocity is produced by the electron-gun at G. This is similar in principle to the gun in a television picture tube and consists of a heated cathode which emits electrons that are accelerated past a series of anodes. The gas or vapour to be studied is injected through a fine nozzle, N, and is condensed out on a cold finger, C. The whole chamber is continuously evacuated to a very low pressure and must be carefully screened from extraneous electric or magnetic fields. In the region between N and C, the electrons are diffracted from the molecules and the diffracted beams are recorded on a fluorescent screen or photographic plate at S. The injection and condensing arrangement is designed to limit the gas or vapour molecules to a narrow trajectory, so that they do not fill the rest of the apparatus. The diffracted intensity falls off very fast with increasing θ, so that a rotating sector must sometimes be placed in front of S to produce differential exposures increasing towards the edges of the plate.

We saw in Chapter 5 that information was lost from the diffraction pattern of randomly-oriented crystallites, compared to what can be observed from a single crystal, even though individual crystallites may be perfect in themselves. In glasses, all ordering is lost but diffuse diffraction rings are produced from the distribution of bond-lengths. In gases and vapours we might expect

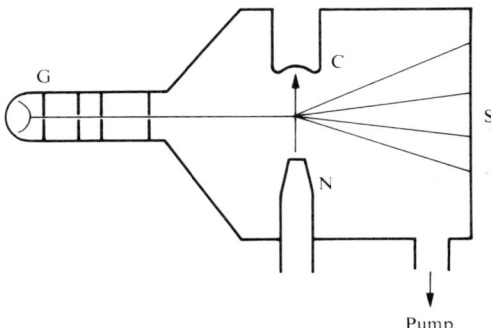

FIG. 7.1. Schematic view of the Wierl apparatus for electron diffraction from gases and vapours.

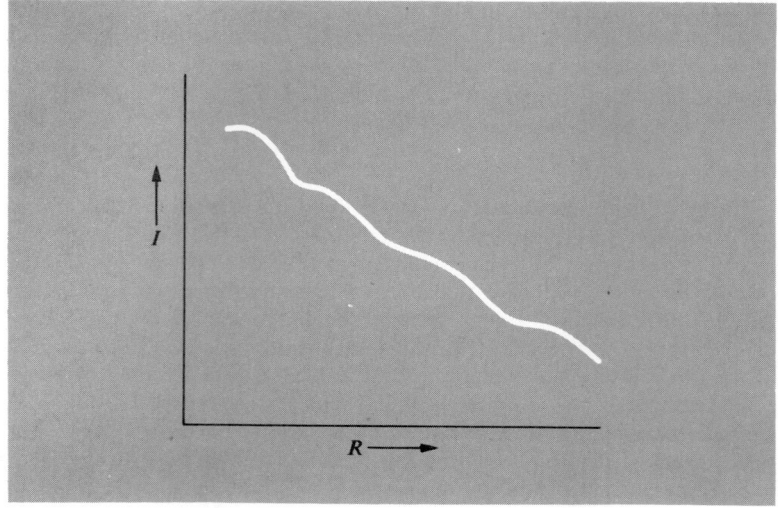

FIG. 7.2. Scattering intensity plotted against distance from the centre of the plate.

nothing at all, but in actual fact, for all but monatomic gases, there is a much more precisely-defined set of bond-lengths. Because of the low concentration, intermolecular diffraction effects become very weak but intramolecular diffraction still remains, on the model of Fig. 1.2. All that happens is that millions of different patterns from all the molecules in random orientations are superimposed. By summing the contributions, the average intensity as a function of angle, $I(\theta)$, is obtained in terms of f_e and the interatomic distance, r, in eqn (7.1) (see Moore (1963)). The general form of the intensity distribution

$$I(\theta) = 2f_e^2(1 + \sin sr/sr) \quad \text{where} \quad s = (4\pi/\lambda)\sin(\theta/2) \qquad (7.1)$$

is plotted against R, the distance from the centre of the plate, in Fig. 7.2. Note the dominating steep slope, representing the rapid decrease in incoherent scattering from individual atoms. Superimposed on this are the small modulations, from cooperative scattering and diffraction between the different atoms. For more complex molecules, the *Wierl equation*, eqn (7.2), is used to calculate the coherent-diffraction intensity, with the summations carried out over all

$$I(\theta) = \sum_i \sum_j f_i f_j \sin(sr_{ij})/sr_{ij} \qquad (7.2)$$

atoms. The observed intensity curve, after removal of the strong background, and the calculated curve are shown for ferrocene (p. 84) vapour at 140°C in Fig. 7.3 (upper), as A and B, respectively. The diffraction curve is in fact the

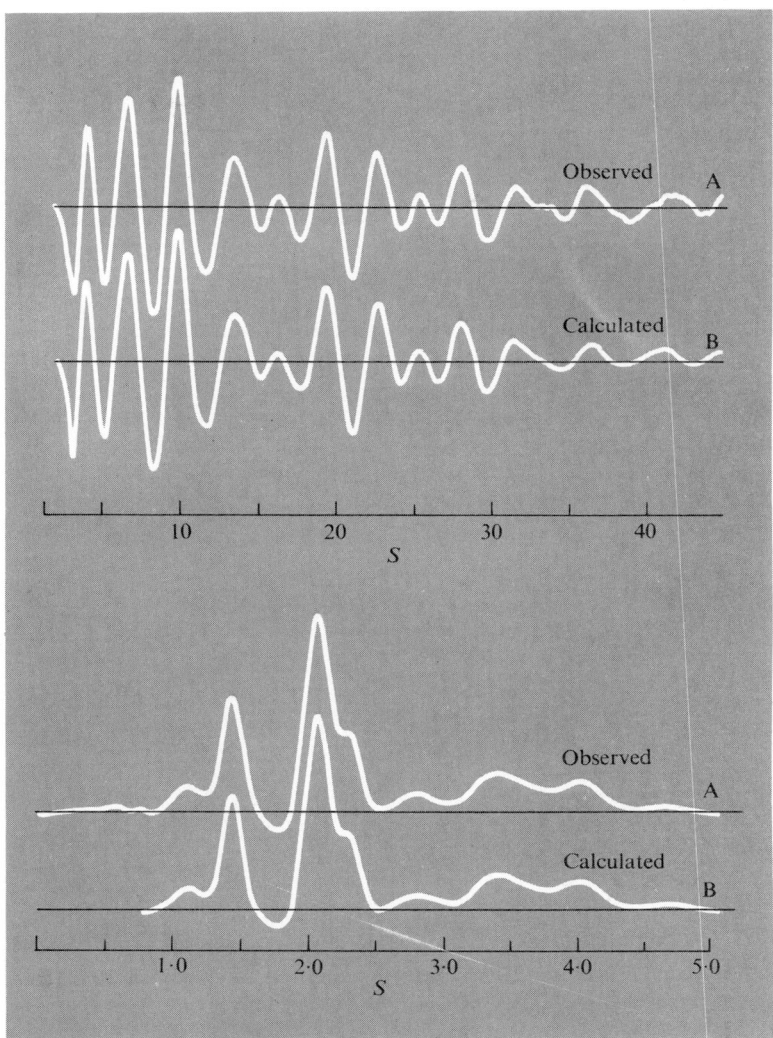

Fig. 7.3. The modified molecular intensity and radial distribution curves for electron diffraction from ferrocene vapour at 140°C.

Fourier transform of the radial distribution of scattering power, for which the observed and calculated distributions are shown in Fig. 7.3 (lower). Note that f_i and f_j can be modified to allow for internal vibrations of the molecule, as in solid-state work.

With such limited data, there is no question of 'solving' structures. Rather, an adequately-accurate model must be available to start with, or a limited choice of alternatives; it is quite possible to force-fit a wrong model to the data, as for $HMn(CO)_5$. Very accurate refinement of the dimensional parameters is then possible, and the bulk of our detailed knowledge of molecular structures in the vapour or gas phase comes from electron-diffraction studies. Admittedly, more precise parameters can be extracted from microwave spectra, but interpretation of these is limited to rather simple molecules with relatively high symmetry. Interesting comparisons can be drawn between the structural data derived from the three main diffraction techniques: the structure of ferrocene is a good example.

Single-crystal X-ray work established structure (12) for ferrocene in the crystal. The π-cyclopentadienyl ligands are planar, parallel, and rotationally staggered by the 'ideal' angle of $360/10 = 36°$. Ruthenocene, which is isoelectronic with ferrocene in the bonding orbitals, has the eclipsed structure (13),

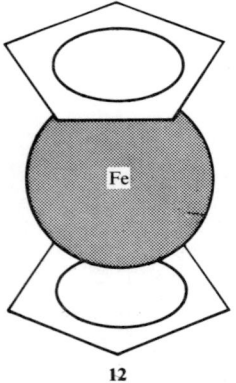

12

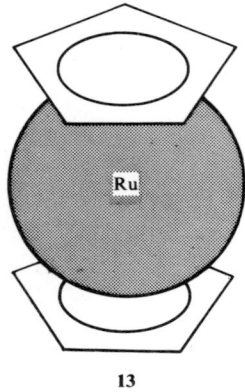

13

as does osmocene. It was therefore quite commonly assumed that there was a genuine chemical distinction between the two types of structure, presumably linked to the different perpendicular distances between the planes of the two ring ligands, 332 pm for (12) and 369 pm for (13), although the original crystallographic paper on (12) did warn against any such assumption. The general idea that interannular repulsions governed the choice between a staggered or eclipsed structure was undermined by many later studies of substituted ferrocenes, which showed a variety of configurations ranging from one extreme to the other, apparently as a balance between interannular and intermolecular packing interactions.

A neutron-diffraction study (Willis, 1960) supported structure (12) and indicated that the hydrogen atoms lay in the planes of their respective ligand rings. This was unfortunately but a preliminary abstract and the fully-refined structure does not appear to have been published. The question of the hydrogen-atom positions is of interest because the free π-cyclopentadienyl ligand might be expected to show strict planarity on account of the sp^2 hybridization of its carbon atoms, whereas this precise hybridization ought to be perturbed by the bonding of the ligand to the iron atom. The latest and most accurate electron-diffraction study (Bohn and Haaland, 1966) provides extremely precise molecular dimensions for ferrocene, with the carbon–carbon bond lengths measured to ± 0.5 pm. It also produces proof that the hydrogen atoms are displaced out of the C_5-ring planes,[†] so that the C—H bonds bend inwards by about 5° towards the iron atom. Most surprisingly, the configuration is an exactly eclipsed one, with an activation energy of $4.6\,\mathrm{kJ\,mole^{-1}}$ opposing internal rotation. This is an interesting example of how differently a molecule can behave in a 'dilute' system, in which it cannot interact significantly with its neighbours, and in a condensed phase. In any theoretical discussion of the bonding in ferrocene, therefore, it is advisable to refer to the electron-diffraction structure.

High-energy electron diffraction from solids

High-energy electron diffraction (HEED) was originally demonstrated by Thomson and Reid (1927), who obtained a transmission diffraction pattern from a thin gold foil, using 10–60 keV electrons. At such energies the electrons can penetrate several microns into solid materials. As originally developed, HEED cameras resembled the Wierl apparatus, but with a suitably prepared and mounted solid specimen replacing the stream of vapour molecules. In modern practice, the facilities offered by the electron microscope are normally used. To begin with, this instrument provides well-developed high-vacuum and sample-handling techniques. Secondly, it is equipped to focus electron beams; this allows a great gain in intensity, since a non-collimated source can be used. HEED and electron microscopy have in fact become very closely associated and the complementary natures of the two techniques are schematically described in Fig. 7.4. Electrons from the source are condensed by the lens C and illuminate the specimen S, shown here as a two-point scatterer for simplicity.

In the microscope (a) the objective lens O recombines different beams scattered from the *same* point in S to form an image in the plane I. This is a magnified, real, image which can be displayed on a fluorescent screen or

† See also McLauchlan (1972), *Magnetic resonance* (OCS 1).

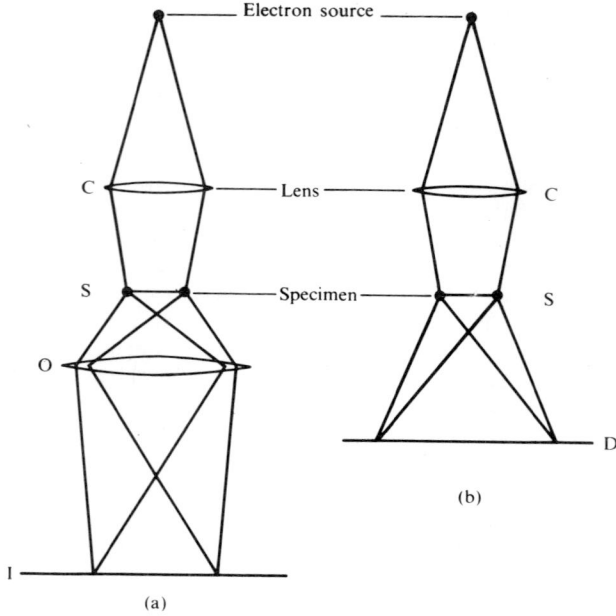

Fɪɢ. 7.4. Diagrammatic representation of the relationship between (a) electron microscopy and (b) HEED.

recorded on a photographic plate. In the HEED mode (b) scattered beams from *different* parts of S are observed at their point of interference, creating a diffraction pattern in the plane D.

A real electron microscope contains far more complex lens systems: these lenses actually consist of electromagnetic coils surrounding the vacuum chamber, whose flux can be adjusted at will. The flexibility of the arrangement allows a number of experimental variations in HEED. For example, an electron micrograph of the sample can be directly superimposed on the diffraction pattern, without disturbing the sample. This is convenient for relating the unit-cell axes to external morphology, especially in the study of textures. The judicious use of screens also allows the diffraction patterns of selected areas of the specimen to be examined, without too great a loss in intensity, which can be valuable with inhomogeneous samples. In some apparatus the electron beam can be made to scan the specimen under electromagnetic control. Lastly, the diffraction patterns of samples may be obtained both in transmission and in reflection.

Since 50 to 100 keV electrons are normally used, the specimens must be very thin. They are usually prepared for transmission studies by microtoming or epitaxial growth. The latter is a technique whereby a crystal of the sample material is grown on a well-defined face of a host crystal of another substance. The host is then either stripped or dissolved away. Very thin yet well-ordered films can be grown in this way. Such thin specimens are very fragile and great care must be exercised in handling them, particularly when they are destined for reflection studies, in order to avoid distorting their lattices. Sample-handling restrictions are one of the reasons why HEED does not compete directly with X-ray diffraction methods. Another problem is that values of f_e for the various elements are known with less accuracy than f_n or f_x. The main value of HEED is therefore not in complete structure determination so much as in the examination of polycrystalline specimens and textures, particularly in conjunction with electron microscopy. The diffraction principle is also used in *phase-contrast microscopy*, which is useful for studying crystal defects. The reader is referred to textbooks on electron microscopy, such as that of Hirsch, *et al.* (1965) for details.

Low-energy electron diffraction

Davisson and Germer (1927) used low-energy electrons, of some 30 to 600 eV, in their original experiment and obtained reflection photographs. The low penetrating power of such electrons at first limited interest in LEED methods but they have recently been revived for the specific purpose of examining the few layers of atoms near the surfaces of specimens. The experimental arrangements are quite simple and are depicted in Fig. 7.5. The gun G produces an electron beam of the desired energy, which enters the evacuated

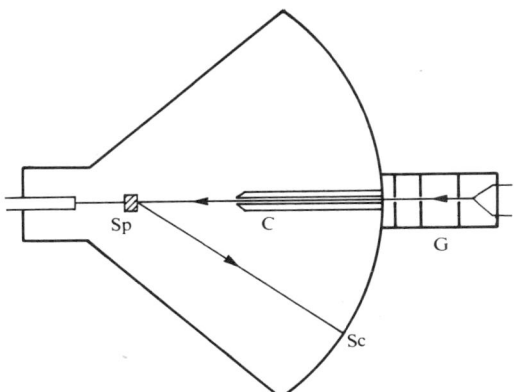

FIG. 7.5. Diagrammatic view of a LEED apparatus.

apparatus through the collimator C and strikes the specimen Sp. Diffracted beams are received on the screen Sc, which is normally given an anodic potential to accelerate them to an energy at which they will excite phosphors or photographic plates. Control grids are used to discriminate against stray and inelastically scattered electrons. Alternatively, the screen may be replaced by a fixed detector; reciprocal space is then scanned by varying λ through control of the primary-beam energy, and rotating the specimen about an axis collinear with the collimator.

Surfaces of crystals corresponding to planes of low Miller indices show the highest symmetry and are the easiest to examine. The three main areas of interest for LEED work are metals, non-metals and compounds, and adsorbed species. In order to examine any of them, one must produce clean surfaces and clean in this context implies the removal of all adsorbed species. This requires vacua down to 10^{-9} torr which more than anything else, hindered the exploitation of LEED for many years. Single crystals of metals are cut to expose the desired plane; the surface is mechanically polished and then chemically or electrochemically etched to remove mechanically-induced distortions. The clean surface is then studied by LEED. The coordination of the metal atoms in the surface is of particular interest when the metal catalyses heterogeneous reactions. Gases can be allowed to adsorb on to the hitherto clean sample under controlled conditions; the diffraction pattern of the adsorbed molecules appears as reflections superimposed on those from the clean surface. In favourable cases, the distribution of molecules on the surface can be related to the existence of particularly active adsorption sites. Preferred orientation and dissociation into atoms can be monitored, particularly for such simple gases as H_2, O_2, and N_2. In non-metals and compounds, atoms or ions at the free surface tend to migrate in order to reorder their distorted coordination environments. This reordering gives rise to diffraction maxima corresponding to fractional index values, based on the original crystal lattice. Precise characterization of surfaces is useful in the study of metal-oxide catalysts. The formation of oxides or the presence of other impurities at the boundaries of grains in metals is another area amenable to investigation by LEED.

8. Conclusion

WE HAVE examined the diffraction phenomenon and some of its practical applications to chemistry. These applications have proved invaluable sources of information, particularly on the structures of molecules and crystals. The range of uses is considerable but needs to be put in some perspective. Diffraction methods are, of course, only a part of the wide choice of physical techniques available to the chemist, who must always choose his tools judiciously in tackling a particular problem. They are, as we have seen, intensively and increasingly used, largely because they have become easier, quicker, and cheaper. Will this trend continue? If so, will diffraction methods become even more attractive?

The choice of a physical method in chemistry is nearly always time-constrained. In fact, quite a good analogy can be drawn with photography: a fast-moving object requires short exposures, which require large apertures and fast films, in turn reducing depth of focus and grain resolution. In flash photolysis, species with a lifetime of one nanosecond can be examined—but the examination consists of a very quick scan of part of their absorption or emission spectrum. Each 'snapshot' therefore gives very limited information. The infrared spectrum of a solution can be obtained in 1–2 minutes with normal instruments; in principle, it contains a host of data, covering all kinds of stretching and bending modes of vibrations; in practice, limited information about the presence or absence of certain characteristic chemical groups is normally derived. The solution obviously must not decompose to any great extent during the scan through the spectrum. Crystals used in the collection of diffraction intensities must normally not decompose appreciably in less than a day—the general rule of thumb is to monitor the intensities of one or more standard reflections and discard the crystal if any one of them declines by 15 per cent or more. The snapshot becomes a very elaborate portrait, full of faithful detail, but requires the subject to stay still for rather a long time. Effective 'exposure times' therefore vary over a range of $1:10^{13}$, approximately, with diffraction methods firmly at the long-exposure end of the range.

The interests of chemists can be broadly divided into structure and reactivity. The borderline between the two cannot be drawn with much certainty—every chemical species, however transient, must have a structure, and every element or compound, however reluctant, will undergo some reaction. Diffraction methods are quite clearly limited to the study of relatively stable species, and their importance is related to that of structural chemistry in general. It is safe to say that the earliest valid notions of structure came out of studies on reactivity—the Law of Multiple Proportions is the classic example.

Since 1900, the situation has reversed itself, beginning with the work of Ingold and his school on the mechanisms of organic reactions, interpreted in terms of the structures of reagents and solvents. We can as yet obtain very little direct information on transition states and must make do with knowledge of the starting reagents and the final products, together with any isolable intermediates. The more detailed this knowledge, the better. Prognostications in science are always rather uncertain but it is probably safe to suggest that X-ray crystallography will be increasingly used to characterize reagents, products, and catalysts. As diffractometers and computing become cheaper and chemists more used to using them, crystal structure determinations will in many cases be routinely applied to coordination complexes and organometallic molecules—the trend is actually well under way now. There will be a great increase in work on natural products and new drugs. We shall see many more three-dimensional structures for enzymes and other quite large biological entities, in so far as these can be crystallized. Technical developments are likely to concentrate on photographic or photo-imaging methods of data collection for very large unit cells and less-stable compounds. Some work is already being performed on solid-state reactions in crystals, which are relatively slow. On the computing side, a few small structures have already been solved fully automatically, without human intervention from the input of the data set to the producing of final atomic coordinates, relying on direct methods of phasing.

In the realm of partial ordering, the methods developed for synthetic polymers will increasingly be extended to biology. Results have already been published on the lamellar structure of bleached and unbleached rod photoreceptor membranes, and even on X-ray diffraction from a whole frog's eye. The former data are summarized in Fig. 8.1 (Corless, 1972). Similar studies have been undertaken on muscle contraction.

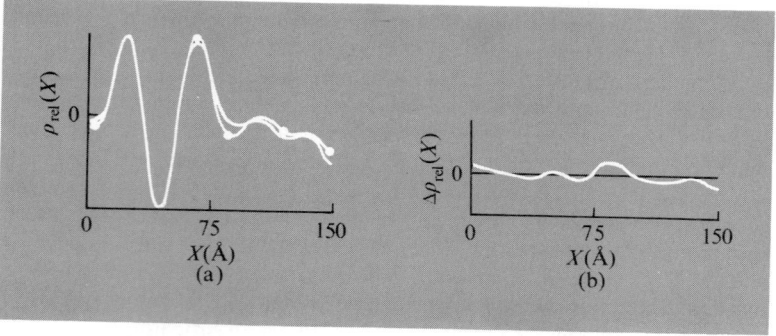

FIG. 8.1. Overlapping-density profiles determined for bleached (————) and unbleached (——●——) disc membranes in (a), and the redistribution produced by bleaching in (b), for the disc membrane of a frog retina (from Corless (1972)).

Much of our chemical thinking is based on structural models, from the simple ionic salts to the most complex enzymes. The real value of diffraction methods is in the models they allow us to construct; through them, we come as close as we can to 'seeing' the structures on whose behaviour chemistry is based.

Appendix I

Crystallographic units of length

LENGTHS in this book are quoted in the SI units of nm and pm. Many crystallographers still use the *ångstrom unit*, or Å, of 10^{-10} m.

$$1 \text{ ångstrom} = 100 \text{ pm} = 0.1 \text{ nm}$$

In some old papers, the kX unit may be encountered; it differs insignificantly from the ångstrom.

Appendix II

The structure-factor equation

EQUATION (2.3) has the form of eqn (II.1);

$$F_{hkl} = \sum_j f_j \exp(B \sin^2 \theta / \lambda^2) \exp(2\pi i(hx_j + ky_j + lz_j)) \tag{II.1}$$

in which the general notation of complex algebra, eqn (II.2);

$$\exp(i\theta) = \cos\theta + i\sin\theta \tag{II.2}$$

is used.

Fourier Series and the Fourier Transform

Fourier's theorem states that any function $F(x)$ can be represented as an infinite series of terms, as in eqn (II.3)

$$\mathfrak{F}(x) = a_0/2 + \sum_{n=1}^{\infty} (a_n \cos nx + b_n \sin nx). \tag{II.3}$$

Electron density in the unit cell can thus be represented by eqn (II.4). Now a

$$\rho(x, y, z) = \sum_{h'=-\infty}^{\infty} \sum_{k'=-\infty}^{\infty} \sum_{l'=-\infty}^{\infty} \{C(h'k'l') \exp(-2\pi i(h'x + k'y + l'z))\} \tag{II.4}$$

structure factor based on scattering by the infinitesimal elements of electron density is given by eqn (II.5)

$$F_{hkl} = V \int_0^1 \int_0^1 \int_0^1 \rho(x, y, z) \exp(2\pi i(hx + ky + lz)) \, dx \, dy \, dz. \tag{II.5}$$

Substituting (II.4) into (II.5) gives eqn (II.6), in which the integrals vanish unless $h = -h'$, $k = -k'$, $l = -l'$, to yield eqn (II.7). Putting (II.7) back

$$F_{hkl} = \int_0^1 \int_0^1 \int_0^1 \sum_{-\infty}^{\infty} \sum_{-\infty}^{\infty} \sum_{-\infty}^{\infty} C(h'k'l') \exp(-2\pi i(h'x + k'y + l'z))$$

$$. \exp(2\pi i(hx + ky + lz))V \, dx \, dy \, dz \tag{II.6}$$

$$F_{hkl} = \int_0^1 \int_0^1 \int_0^1 C(h'k'l')V \, dx \, dy \, dz = C(h'k'l')V \tag{II.7}$$

$$\rho(x, y, z) = \frac{1}{V} \sum_{-\infty}^{\infty} \sum_{-\infty}^{\infty} \sum_{-\infty}^{\infty} F_{hkl} \exp(2\pi i(hx + ky + lz)) \tag{II.8}$$

into (II.4) gives eqn (II.8), which is equivalent to eqn (2.5) with one small exception. We cannot obtain indices from $-\infty$ to $+\infty$ because of the resolution limit: a *series termination error* therefore affects all real Fourier maps. This relation between electron densities and structure factors is a form of *Fourier transform*.

Bibliography

Chapter 1

FRIEDRICH, W., KNIPPING, P., and LAUE, M. (1912). *Proc. bav. acad. Sci.*, 303.
LIPSON, H. and COCHRAN, W. (1966). *The determination of crystal structures*, Fig. 17 (opp. p. 17). Cornell University Press, Ithaca, N.Y.

Chapter 2

BRAGG, W. L. (1913). *Proc. Camb. phil. Soc.*, **17**, 43–57.
International tables for X-ray crystallography (1965), vol. 1, (ed. N. F. M. Henry and K. Lonsdale). The Kynoch Press, Birmingham.
MONTEATH ROBERTSON, J. (1936). *J. chem. Soc.*, 1195.

Suggestions for further reading

BUERGER, M. J. (1942). *X-Ray crystallography*. Wiley, New York, pp. 60–90. A description of symmetry elements in space groups and their effects on diffraction patterns.
BURKE, J. G. (1966). *Origins of the science of crystals*. University of California Press. An historical account of the development of crystallography.
COTTON, F. A. (1966). *Chemical applications of group theory*. Academic Press, New York, pp. 14–49. A description of symmetry elements and molecular point groups.
HITCHCOCK, P. B. and MASON, R. (1971). Direct solution of crystal structures—the phase problem, *Chemistry in Britain*, **7** (December), p. 511.

Chapter 3

CHURCHILL, M. R. and WORMALD, J. (1971). *J. am. chem. Soc.*, **93**, 5670.
CHURCHILL, M. R. WORMALD, J., KNIGHT, J., and MAYS, M. J. (1971). *J. am. chem. Soc.*, **93**, 3703.
CHURCHILL, M. R., SCHOLER, F. R., and WORMALD, J. (1971), *J. organomet. Chem. (Amsterdam)*, **28**, C21.

Suggestions for further reading

ARNDT, U. W. and WILLIS, B. T. M. (1966). *Single crystal diffractometry*. Cambridge University Press. A readable account of the design philosophy behind diffractometers.
BUERGER, M. J. (1970). *Contemporary crystallography*. McGraw-Hill, New York. This book contains particularly good reproductions of X-ray diffraction photographs, together with explanations of the various cameras, by a pioneer of their design.
BUERGER, M. J. (1960). *Crystal structure analysis*. Wiley, New York. Gives a very thorough account of the geometrical basis of diffraction from crystals. Definitive but quite difficult. Buerger has written a series of such textbooks, covering most aspects of X-ray crystallography. His particular field is the study of minerals.
STOUT, G. H. and JENSEN, L. H. (1968). *X-Ray structure determination: a practical guide*. Macmillan, London. The best attempt so far to produce an A-to-Z guide to solving a crystal structure. Readable, complete, and highly recommended.

Chapter 4

ADDISON, W. E. (1961). *Structural principles in inorganic compounds.* pp. 81–83, Longmans, London.

COULSON, C. A. (1961), *Valence* (2nd edn.). pp. 301–319, Oxford University Press, London.

FLEISCHER, E. B. (1972). *J. am. Chem. Soc.*, **94**, 1382.

FLEISCHER, E. B. (1970). The structure of porphyrins and metalloporphyrins, *Acc. chem. Res.*, **3**, 105–112.

KENNARD, O. and WATSON, D. G. (1970). *Molecular structures and dimensions*, vol. 1. IUC/NVA Oosthoek's Uitgevers Mij, Utrecht.

MASON, R. and ROBERTSON, G. B. (1966). In *Advances in structure research by diffraction methods*, vol. 2 (ed. R. Brill and R. Mason). Vieweg/Interscience, pp. 35–74.

WELLS, A. F. (1962). *Structural inorganic chemistry* (3rd edn.). Clarendon Press, Oxford.

WELLS, A. F. (1956). *The third dimension in chemistry*, pp. 87–133. Oxford University Press, London.

WYCKOFF, R. G. (1967). In *Werner Centennial* (ed. G. B. Kauffman), pp. 114–119. American Chemical Society, Washington, D.C.

Suggestions for further reading

DICKERSON, R. E. (1972). *The structure and history of an ancient protein, Scientific American (April)*, 58–72.

KARTHA, G. (1968). *Picture of Proteins by X-ray diffraction*, *Acc. chem. Res.*, **1**, 374.

MARRISON, L. W. (1966). *Crystals, diamonds, and transistors.* Penguin Books, Harmonsworth, Middlesex.

Chapter 5

ALEXANDER, L. E. (1969) *X-ray diffraction methods in polymer science.* Wiley-Interscience, New York.

AZAROFF, L. V. and BUERGER, M. J. (1958). *The powder method in X-ray crystallography.* McGraw-Hill, New York.

URNES, S. (1960). X-ray diffraction studies of glass, in *Modern aspects of the vitreous state* (ed. J. D. Mackenzie). pp. 10–37, Butterworths, London.

Suggestions for further reading

BRADY, G. W. (1971). Some aspects of small-angle X-ray scattering, *Acc. chem. Res.*, **4**, 367–373.

NYBURG, S. C. (1961). *X-ray analysis of organic structures.* pp. 325–396, Academic Press. Good accounts of structural X-ray work on macromolecules and fibres.

WATSON, J. D. (1968). *The double helix.* Weidenfeld and Nicolson, London. How the structure of DNA was elucidated; a highly personal inside view.

Chapter 6

BACON, G. E. (1964). The determination of crystal structures by neutron diffraction measurements. In vol. 1 of *Advances in structure research by diffraction methods* (ed. R. Brill). pp. 1–23, Vieweg/Interscience.

BACON, G. E. (1966b). The determination of magnetic structures by neutron diffraction. In vol. 2 of *Advances in structure research by diffraction methods* (ed. R. Brill and R. Mason). pp. 1–34, Vieweg/Interscience.

BARRETT, C. S. and MASSALSKI, T. B. (1966). *Structure of metals*, p. 608. McGraw-Hill, New York.

CHURCHILL, M. R. and WORMALD, J. (1971). *J. am. chem. Soc.*, **93**, 5670.

COULSON, C. A. (1961). *Valence*, ch. 13, pp. 344–356. Oxford University Press, London.

HIEBER, W. and WAGNER, G. (1958). *Z. Naturf.*, **13b**, 340.

IBERS, J. A. (1964). *Ann. rev. Phys. Chem.*, **16**, 375.

LAPLACA, S. J., HAMILTON, W. C., IBERS, J. A., and DAVISON, A. (1969). *Inorg. chem.*, **8,**.1928.

PETERSON, S. W. and LEVY, H. A. (1957). *Acta crystallogr.*, **10**, 70.

Suggestions for further reading

BACON, G. E. (1966a). *X-Ray and neutron diffraction*, Pergamon Press, Oxford. A general introduction to the subject, based on a selection of historic original papers.

Chapter 7

BOHN, R. K. and HAALAND, A. (1966). *J. organomet. chem.* (*Amsterdam*), **5**, 470.

DAVISSON, C. J., and GERMER, L. H. (1927), *Phys. Rev.*, **30**, 705.

MOORE, W. J. (1963). *Physical chemistry* (4th ed.), pp. 575–580, Longmans, London.

THOMSON, G. P. and REID, A. (1927). *Nature*, **119**, 180.

VAINSHTEIN, B. K. (1964). Fourier synthesis of potential in electron diffraction structure analysis and its application to the study of hydrogen atoms. In *Advances in structure research by diffraction methods*, vol. 1 (ed. R. Brill), Vieweg/Interscience.

WILLIS, B. T. M. (1960). *Acta crystallogr.*, **13**, 1088.

Suggestions for further reading

RYMER, T. B. (1970). *Electron diffraction*, Methuen, London. A useful general introduction, covering most aspects of the subject.

HIRSCH, P. B., HOWIE, A., NICHOLSON, R. B., PASHLEY, D. W., and WHELAN, M. J. (1965). *Electron microscopy of thin crystals*, Butterworths, London.

LANDER, J. J. (1965), Low-energy electron diffraction and surface structural chemistry, *Progr. solid-state Chem.* (H. Reiss, ed.), **2**, Pergamon Press, Oxford.

Chapter 8

CORLESS, J. M. (1972). *Nature, Lond.*, **237**, 229.

Answers to problems

2.2. Only 1-, 2-, 3-, 4-, and 6-fold axes of rotation are compatible with lattices. You can convince yourself, for example, that rectangles, equilateral triangles, squares, and regular hexagons can be fitted together in infinite two-dimensional arrays, but not regular pentagons.

2.4.

$$2x, 2y, 2z \quad (1)$$

$$2x, -2y, 2z \quad (1)$$

$$0, \tfrac{1}{2} \pm 2y, \tfrac{1}{2} \quad (2)$$

$$\pm 2x, \tfrac{1}{2}, \tfrac{1}{2} \pm 2z \quad (2)$$

The relative heights are in parentheses. To find these, one subtracts one symmetry position from another, taking all pairs in turn, remembering that the Patterson map is inherently centrosymmetric and that ± 1 can be added to any coordinate, since any one unit cell is like its neighbours.

3.1. They give values for a^*, b^*, c^*, and β^* but not for α^* and γ^*.

3.3.

$$I = kx^3 I_0 \exp(-\mu x)$$

$$\mathrm{d}I/\mathrm{d}x = 0 = 3kx^2 I_0 \exp(-\mu x) - \mu k x^3 I_0 \exp(-\mu x)$$

$$3x^2 - \mu x^3 = 0$$

$$x = 3/\mu$$

This enables you to choose optimum crystal-dimensions—a sphere of diameter $3/\mu$ would be best.

4.2. Test refinements with a model having ordered ammonium ions (preferably with neutron diffraction data) will allow you to observe the effects of temperature on the apparent thermal-vibration parameters of the hydrogen atoms.

4.3. A non-centrosymmetric space group. The relationship $(x, y, z) \equiv (\bar{x}, \bar{y}, \bar{z})$ would require a racemic mixture to be present.

4.5. The unit-cell dimensions are effectively doubled. Indexing on the original cell, reflections $(h/2, k/2, l/2)$ will appear.

5.2. Only these elements have unpaired electrons buried in inner orbitals (3d, 4d, 5d, 4f, and 5f).

5.3. Coherent-diffraction losses from the primary X-ray beam are small relative to absorption losses; the reverse is true for neutrons.

Index

absorption coefficient, 27
absorption correction, 28
antiferromagnetism, 76
asymmetric unit, 13
azulenetriruthenium heptacarbonyl, 32

back-donation, 46
band theory of metals and
  semiconductors, 50
Bijvoet pairs, 51
bisfulvalenedi-iron, 30
Born–Haber cycle, 37
Bragg's Law, 7
Bravais lattices, 9

caesium chloride structure, 37
carboranes, 44
catalytic activity of complexes, 43
coordination compounds, 42
coordination number, 37
copper cluster compound, 31
covalent radii, 41
crystal field theory, 42
cytochrome c, 52

de Broglie equation, 6
Debye–Scherrer cameras, 55
degree of crystallinity, 59
difference Fourier maps, 18
diffraction condition, 3
diffraction from an array of points, 4
diffraction from a dipole, 3
diffraction from lattices, 11
diffraction from molecular liquids, 67
diffractometers, 25
diffuse and discrete scattering, 66
direct methods, 16

elastic scattering of neutrons, 69
electron form factor, 80
electron microscopy, 85
estimated standard deviations, 33

ferrimagnetism, 78
ferrocene structure, 84
ferromagnetism, 75
fibres, 59, 63
Fourier maps, 17
fractional occupancy of sites, 70

Fresnel experiment, 2

glasses, 59, 67
growth of X-ray crystallography, 35

high polymers, 58
hydride ligands, 72
hydrogen bonding, 70

inelastic neutron scattering, 79
intensity data, 24
intensity data from films, 24
interstitial compounds, 50
interstitial solid solutions, 50
ionic radii, 38
isomorphous replacement, 16

K_α characteristic peaks, 20

layer lines, 21
least-squares refinement, 18
limiting sphere, 12
liquid crystals, 12
liquid crystals, 67
low-energy electron diffraction from
  surfaces, 88
LP correction, 27

magnetic domains, 77
magnetic form-factor, 74
magnetic scattering, 74
magnetic superlattice, 77
magnetic unit cell, 77
micro- and macro-structure in
  polymers, 64
Miller indices, 11
molecular orbital theory, 42
mosaic crystals, 27

nuclear reactors, 68

organometallic compounds, 45
orientation in polymers, 63
oscillation photograph, 21

packing in crystals, 8
paramagnetism, 74
Patterson maps, 15
pentacarbonylmanganese hydride, 72

phase diagrams in alloys, 58
phase problem, 15
point groups, 9
polarized neutron beams, 74
polythene, 65
powder diffraction, 55
precession cameras, 24

radius-ratio rule, 38
reciprocal lattice, 12
reciprocal space, 12
refinement of structures, 18
resolution, 12
rotation photograph, 23

silicates, 39
small-angle X-ray diffraction, 66
sodium chloride structure, 37
space groups, 11
sphere of reflection, 12
spinels, 78
stereographic projection, 64
structure-factor calculation, 17
structure-factor equation, 14

structure of D_2O at $-50°C$, 70
substitutional solid solutions, 49
superexchange, 79
symmetry elements, 9
symmetry operations, 11
systematic absences, 24

tacticity in polymers, 64
textures, 66
thermal neutrons, 68
trial and error, 15

unit cell, 8
unit-cell axes, 8

wavefront, 2
Weissenberg photograph, 23
white X-radiation, 20
Wierl apparatus, 81
Wierl equation, 82
Wilson plot, 29

X-ray diffraction from partially-ordered
    biological structures, 90
X-ray form factor, 14